HOME BUILDER'S GUIDE TO CONTINUOUS IMPROVEMENT

Schedule, Quality, Customer Satisfaction, Cost, and Safety

HOME BUILDER'S GUIDE TO CONTINUOUS IMPROVEMENT

Schedule, Quality, Customer Satisfaction, Cost, and Safety

Jack B. ReVelle
Derek N. Margetts

CRC Press is an imprint of the
Taylor & Francis Group, an **informa** business

CRC Press
Taylor & Francis Group
6000 Broken Sound Parkway NW, Suite 300
Boca Raton, FL 33487-2742

Printed in the United States of America on acid-free paper
10 9 8 7 6 5 4 3 2 1

International Standard Book Number: 978-1-4200-5507-8 (Paperback)

Library of Congress Cataloging-in-Publication Data

ReVelle, Jack B.
Home builder's guide to continuous improvement : schedule, quality, customer satisfaction, cost, and safety / authors, Jack B. ReVelle, Derek N. Margetts.
p. cm.
Includes bibliographical references and index.
ISBN 978-1-4200-5507-8 (pbk. : alk. paper)
1. House construction--Quality control. 2. Building--Superintendence. 3. Business logistics. I. Margetts, Derek N. II. Title.

TH4812.R478 2010
690.068--dc22 2009036639

Visit the Taylor & Francis Web site at
http://www.taylorandfrancis.com

and the CRC Press Web site at
http://www.crcpress.com

Contents

Preface

This book is intended to provide a basic reference source for use by home builders who want to use tools, techniques, and methods from various disciplines that are applicable to continuous improvement of builders' production schedule, quality, customer satisfaction, cost, and safety.

When this book was begun, home builders were truly enjoying the fruits of their labors; a multi-year streak of record-breaking new home sales and profits (2003–2005). However, as the book approaches completion and the time has arrived when the manuscript is being sent to our publisher, the streak is long gone, almost forgotten, and the residential construction industry is experiencing one really painful year after another (2007–2009). Now, as this book goes to press, American home builders are just beginning to sense that sales activity in the residential market is starting to come alive again.

The coming together of this resurgence in home sales and the publication of this book couldn't have been better timed. It's not enough to simply survive the housing collapse; in order to compete and flourish in the recovery years given the new economic paradigm, pricing, and customer expectations, it is vital for serious homebuilders to maximize their use of those tools that will reduce their cycle times and defect rates as well as their operational and administrative costs and, simultaneously, increase their returns on investment. There is no better way to take advantage of today's situation than working leaner and faster, with greater quality and customer satisfaction levels, with lower costs and accident rates, in order to come out swinging and obtain your piece of the post-recession pie.

This book provides home builders, no matter how large or small, with not only complete descriptions of the tools and techniques, but also examples of their applications in language that is easy to understand. The tools and techniques of continuous improvement can be applied to any critical process in every important aspect of the business (i.e., schedule, quality, customer satisfaction, cost, and safety), and are virtually guaranteed to help serious builders to turn it around.

No matter which functions, departments, or processes in your company need help in making continuous improvement happen, we encourage you to get started now to make a difference in the way you build, the way you work with your trades, suppliers, and contractors, and how you support your home buyers, the customers who make your business succeed.

We wish you a long and continuing series of successful continuous improvement projects. Your feedback to us regarding the usefulness of the book and its value to your company will always be appreciated.

In addition to the examples contained in this book to illustrate the use and impact of various continuous improvement tools, we have prepared a number of case studies from actual homebuilders and trade contractors to demonstrate the real impact and achievements that have been realized using the tools described in this book. These supplemental materials are available to our readers at no cost for review and download at the CRC Press Web site at http://www.crcpress.com/product/isbn/9780750309721. We encourage you to take advantage of these valuable materials intended to help you achieve real, measurable, and sustainable improvement.

Between us, we have over two decades of residential construction experience. As a result of our exposure to both the many great years enjoyed by home builders and the last few dismal years that put many builders out of business and severely impacted those who still remain in business, we strongly believe that now more than ever the industry must push itself to improve product and service quality, schedule performance, and customer satisfaction in order to ride out the storm and make a full recovery possible.

We'd like to express our appreciation to several people who deserve special recognition for their exceptional contributions to the preparation of this book. To Shayna Murry, our sincere appreciation for the excellent layout, color selection, and artwork she provided for the front and back covers. To Karen Simon, our project editor, whose calm and cool kept us on the straight and narrow while we were moving from manuscript to page proofs and beyond. To Maria Muto-Porter who created the majority of the graphics and finalized our manuscript with her consistent professional touch, we value your assistance and patience as we continued to introduce enhanced graphics and verbiage.

Jack B. ReVelle, Ph.D.
A Consulting Statistician
ReVelle Solutions, LLC
Santa Ana, California
Derek N. Margetts
Director of Market
Research & Analytics
Phoenix, Arizona

Authors

JACK B. REVELLE

Since 1979, Dr. Jack B. ReVelle has provided his advice and assistance as a consulting statistician to manufacturing, service, and residential construction organizations throughout North America, Britain, and China. In this capacity, he helps his clients to locate and significantly reduce their process pain. This pain includes excess cycle time duration and variation, excess labor and material costs, excess defects in production and errors in administration, excess process complexity, insufficient process documentation and unacceptably low customer satisfaction scores. Better understanding and continuous improvement of his client processes are accomplished through the use of a broad range of Six Sigma–related tools, techniques, and methods. These include, but are not limited to: process mapping, statistical sampling, root cause analysis, strategic planning, quality function deployment, design of experiments, statistical process control, as well as integrated product and process development. Dr. ReVelle also provides his technical assistance as both a quality and an industrial engineering expert to attorneys involved in litigation. Dr. ReVelle was the 2007–2008 chair of Orange County SCORE-114.

Previously, he was the Director of the Center for Process Improvement for GenCorp Aerojet in Azusa and Sacramento, California where he provided technical leadership for the Operational Excellence program. This included support for all Six Sigma, Lean/Agile Enterprise, Supply Chain Management, and High Performance Workplace activities. Prior to this, he was the Leader of Continuous Improvement for Raytheon (formerly Hughes) Missile Systems Company in Tucson, Arizona. During this period, he established and led the Hughes team that won the 1994 Arizona Pioneer Award for Quality and the 1997 Arizona Governor's Award for Quality. He also established the Hughes team responsible for obtaining ISO 9001 registration in 1996.

Dr. ReVelle's previous assignments with Hughes Electronics were at the corporate offices as the Chief Statistician (Director, Statistical and Process Improvement Methods) and Manager, Employee Opinion Research and Training Program Development. Prior to joining Hughes, he was the Founding Dean of the School of Business and Management at Chapman University in Orange, California and Founding Chairman of the Decision Sciences Department at the University of Nebraska in Omaha, Nebraska.

Dr. ReVelle has been inducted into the Purdue University ROTC Hall of Fame (2006) and was awarded the Oklahoma State University College of Engineering, Architecture and Technology (CEAT) Lohmann Medal (2006). He is a three-time Fellow having been elected by: the Institute of Industrial Engineers (1993), the American Society for Quality (1992), and the Institute for the Advancement of Engineering (1987). He is listed in *Who's Who in Science and Engineering, Who's Who in America, Who's Who in the World,* and as an Outstanding Educator in *The International Who's Who in Quality.*

Dr. ReVelle is a 1990 recipient of the Distinguished Economics Development Programs Award from the Society of Manufacturing Engineers, a 1991 recipient of the Taguchi Recognition Award from the American Supplier Institute, a 1999 recipient of the Akao Prize from the Quality Function Deployment (QFD) Institute, and a 1999 recipient of the Distinguished Faculty Award from The National Graduate School of Quality Management. He is one of only two persons in the world who has been honored to receive both the *Taguchi Recognition Award* for his successful application of Robust Design and the *Akao Prize* for his outstanding contribution to the advancement of QFD.

Dr. ReVelle received his B.S. in chemical engineering from Purdue University, and both his M.S. and Ph.D. in industrial engineering and management from Oklahoma State University. Prior to receiving his Ph.D., he served 12 years in the U.S. Air Force. During that time, he was promoted to the rank of major and was awarded the Bronze Star Medal while stationed in the Republic of Vietnam as well as the Joint Service Commendation Medal for his work in quality assurance with the Defense Nuclear Agency.

DEREK N. MARGETTS

Derek Margetts has been involved in the residential construction industry since 1998. He currently serves as director of research and consumer analytics for a large, privately held home builder. In this position, he is responsible for understanding the wants and needs of the consumer and turning that data into knowledge that can be acted upon to produce results. Previously he has held positions in construction as an operations analyst where he was involved in process improvement projects ranging from cycle time reduction to new process development, quality assurance, and cost analysis. He received his B.S. in psychology in 2000 and his M.B.A. in 2005, both from Arizona State University. As an undergraduate, he focused on social influence and cognitive psychology. During this time he participated in several faculty-led research projects and has applied that knowledge and experience to derive models of customer satisfaction in residential construction. As part of his M.B.A. studies, Margetts focused on applied statistics in industrial engineering. This included design of experiments (DOE), statistical process control (SPC), regression models, and nonparametric statistics.

1 Introduction

1.1 PURPOSE OF BOOK

Where does the home builder who cares about continuous improvement of his or her critical processes go for help? Researching this topic for several hours on the National Association of Home Builders (NAHB) bookstore website as well as on Amazon, Barnes & Noble, and the like did not reveal any such books.

There is no direct competition to this book, but as you can imagine, there are many books about the various aspects of home building and many others about continuous improvement. However, none specifically offer insights into continuous improvement of all residential construction-related processes that directly contribute to successful home building.

Look around, and you'll see there isn't a single book available to provide the important insights and information necessary to help reduce cycle time duration and variation, to improve quality and customer satisfaction, and to minimize costs and accidents—until now.

Based on the authors' experiences, described later in this chapter, there is no question that, for whatever the varying reasons may be, virtually every home builder, whatever its size, would like to see its critical process cycle time duration and variation reduced. Reduction of cycle time duration serves several purposes; for example, it allows more homes to be built without benefit of additional resources and reduces the interest costs paid on loans used to purchase the land upon which the homes are being built.

Reduction of cycle time variation also results in more consistency in construction schedules. This simplifies the scheduling of the builders' trades and subcontractors that, in turn, are better able to plan the allocation of their personnel and equipment resources. Thus, their costs are reduced, which allows them either to reduce their invoices to builders, to earn greater profits, or both.

Improving both construction quality and customer satisfaction also pay excellent dividends. For example, improved construction quality means fewer defects as well as errors of both omission and commission. This results in fewer customer complaints, greater customer satisfaction, and ultimately, more referrals. Every builder either already knows (or should know) that referrals are the least expensive and most productive form of marketing.

Minimizing the costs of construction, sales and marketing, landscaping, options, land acquisition, land development (engineering), design centers, purchasing, finance and administration, and all the other necessary functions associated with a home builder's organization as well as the direct and indirect costs of accidents are (or should be) important to every builder.

This book addresses all these needs. The tools and techniques of continuous improvement are well known throughout general industry, health care, education, military, and even among a few of the giant builders, but the typical home builder is constrained by the lack of availability of time, money, and personnel with the necessary and appropriate knowledge and experience to improve its critical processes. Acquisition of these specialized personnel requires either costly search and recruitment efforts or the expense of training and time away from the job for existing associates.

The discussions that follow draw on both our early and recent experience (one as a consulting statistician and the other as regional home builder executive) and present a logical sequence of topics, each one building (pun intended) on the information introduced in earlier chapters. No longer must the typical home builder lament the lack of information available to help create a sustainable climate of continuous improvement.

More and more of today's home buyers are studying the J.D. Power rankings of home builders in the cities where they plan to purchase new homes. These rankings of customer satisfaction help both new and experienced home buyers to quickly locate the home builders that have been most successful in continuously improving their products, processes, and customer services.

1.2 TARGET AUDIENCE

Many persons will benefit from this book, including the following:

- Residential builder executives, directors, managers, supervisors, and construction superintendents
- Owners and general managers of trade contractors who view continuous improvement of their functions and processes as a major component of their organizational responsibilities
- University faculty and students in construction management programs

1.3 BENEFITS OF THE BOOK

This book responds, in part, to a quote from the May 2006 issue of *Builder* magazine regarding production builders who "say that improving cycle time will be the most important factor in continuing to grow. Last year, the average cycle time for The Builder 100 companies increased to 141.81 days, up from 138.57 in 2004." Now that the 10-plus years of a great seller's market are behind us, this book provides production builders and their trade contractors with the right tools used in the right sequence for the right reasons to reduce cycle time duration and variation, improve quality and customer satisfaction, and minimize costs as well as the frequency and severity of accidents.

The major difference between this book and the various generic books on continuous improvement noted earlier is that all the examples in this book are specific to the authors' personal experiences as a statistical consultant and home builder executive working within the residential construction industry for over 10 years with over

10 national, regional, and local home builders as well as about 20 of their trade contractors. In addition, consider the following:

- There are plenty of figures and graphics for concept clarification.
- The book contains easy-to-follow, nontechnical language using residential construction industry terminology to assist reader understanding of continuous improvement concepts and practices.
- No previous math or statistics background is needed to make maximum use of the graphs or equations in the book.
- The book is designed to be useful to persons at all levels and in all functions of residential builder and trade contractor organizations.

1.4 EXAMPLES

Unlike other books written and published about the judicious application of continuous improvement problem-solving tools and techniques, this book employs residential construction examples to facilitate learning experiences.

Examples provide a unique approach to learning. Rather than simply using theory without a real-world context, examples present an in-depth examination of a particular type of situation. They provide a systematic way of looking at problems, collecting data, analyzing information, and reporting results.

Examples are a documented account of problem identification and resolution activity as it actually occurred, but without identification of the involved companies; they are narratives that include relevant information for consideration by the reader. An example is an edited, real-world situation used to work through particular principles and to assist the reader in understanding and achieving specific learning outcomes.

1.5 BOOK FORMAT

The table of contents reflects the ordering of the chapters in the book. The ordering didn't just happen; it was planned to provide a logical learning sequence that would maximize the reader's intellectual experience.

The authors highly recommend that the reader seriously consider the following advice. As just noted, the book was designed and written to be read, studied, and applied in the chapter order presented. Selecting and using chapters at random, here and there, rather than in the existing sequence, is quite likely to contribute to some lesser level of success other than starting at the front of the book and then progressing to the final thoughts. The authors recognize the reader will use the book as the reader sees fit, but let the reader be warned: the most successful applications resulting from the use of this book will be generated when the reader accepts the authors' recommendation.

Chapter 1, "Introduction," presents the purpose of the book, the target audience, the benefits of the book, an overview of the case studies and why they are used in the book, the authors' construction experience, and the format of the book.

Chapter 2, "Cycle Time Management," exposes the reader to the dual concepts of cycle time duration and cycle time variation. This chapter explains both the basics and advanced steps necessary to abbreviate process duration, to cause process cycle times to be more predictable, and to reduce process complexity through the elimination of nonvalue-added steps.

Chapter 3, "Problems," demonstrates how to identify and define problems, both at construction sites and in the office, as well how to prioritize those same problems. Problems, within the perspective of this book, are situations when processes consistently result in suboptimal results, that is, performance metrics that reflect poorly on the capability and capacity of the processes.

Chapter 4, "Problem Solving," discusses the objectives of logically based problem solving and then introduces four commonly used problem-solving methodologies, all of which employ many similar tools and techniques.

Chapter 5, "Dealing with Data," initially focuses on a variety of quantitative variables and their relationships to each other. Then attention is turned to the nature of various types of data, how data are accumulated, and how they are analyzed. It concludes with a discussion of data patterns, that is, the various types of frequency distribution curves that home builders may have occasion to use in the analysis of their data.

Chapter 6, "Root Causes," reviews the two facets of root cause analysis: cause determination and primary cause selection. One set of tools and techniques is used to determine a range of possible root causes and a completely different set to select the most likely root cause.

Chapter 7, "Corrective Actions," leads the reader beyond identifying and selecting the most likely root cause or causes and then provides an introduction to determining, selecting, and implementing the most appropriate corrective action. The tools necessary to accomplish these objectives are also provided in this chapter.

Chapter 8, "Problem Follow-Up," gives direction regarding what to do after a problem has been identified and the corrective action has been implemented. This includes further data collection and analysis followed by whatever additional process modification is determined to be necessary.

Chapter 9, "Relationships," provides tools and techniques necessary to determine whether two variables are in any way mathematically related and, if so, how to express the predictive relationship between the two in a user-friendly way.

Chapter 10, "Randomization," introduces a concept that ensures every data point has an equal opportunity of being selected for inclusion in a sampling of data drawn from a population. Decisions based on nonrepresentative information can often be incorrect and biased. When 100% inspection or convenience sampling is your current modus operandi, you can be sure that you're either paying too much for the information or using erroneous data or both.

Chapter 11, "Sample Size Determination," discusses the selection of a set of elements from a population of products or services. Sampling is frequently used because population data are often impossible, impractical, or too costly to collect. When this is the case, a sample is used to draw conclusions or make decisions about the population from which the sample is drawn. Users of sampling are cautioned about the three categories of sampling error: bias (lack of accuracy), dispersion (lack of precision), and nonreproducibility (lack of consistency).

Chapter 12, "Software," introduces several "minimum cost, maximum results" software packages designed to provide user-friendly data entry and fast, easy-to-understand results. One package in particular, "QI Macro" was developed by Jay Arthur and is available for purchase via the Internet, the phone, or snail mail. A complementary copy of the software is available for downloading and trying out. The authors have seen this software successfully used by about 10 residential builders and hundreds of their associates.

Chapter 13, "Continuous Improvement Tools and Techniques," explains two dozen tools and techniques, discussed in sufficient detail to ensure that even novice users will find it easy to begin to make use of the right tool or technique to resolve virtually any unacceptable situation or condition. All examples are focused on various facets of residential construction.

Chapter 14, "Quality Function Deployment," provides an overview of the concept of QFD. In the world of business and industry, including residential construction, everyone has customers. Some have only internal customers, some just external customers, and some have both. When a team is working to determine what needs to be accomplished to satisfy or even delight its customers, then the tool of choice is QFD. One of the major reasons QFD is so important in residential construction is the all-too-often failure to communicate between home builders and home buyers. First developed in Japan as a form of cause-and-effect analysis (one of the seven quality control (7-QC) tools in the late 1960s, QFD was brought to the United States in the early 1980s. It gained its early popularity as a result of numerous successes in the automotive industry but has since been successfully applied in scores of industries.

Chapter 15, "Design of Experiments," explains the three major approaches to experimental design and the step-by-step process of DOE. Also included in the chapter are complete discussions of factorial design, orthogonal arrays, and parameter design. The chapter concludes with a detailed checklist for experimentation in a residential construction context. DOEs can be structured to obtain useful information in the most efficient way possible. One intent of applying DOE is to minimize the assets required to obtain a maximum quantity of much-needed data. The assets include cost, time, and physical resources such as capital equipment, raw materials, and test facilities. The language of DOE is influenced by its genesis in the scientific, statistical, and engineering communities.

The book also contains some valuable appendices, which are provided to further assist the reader in understanding the concept and application of continuous improvement. These appendices should support the reader's quest for additional insights regarding continuous improvement of critical processes:

- A comprehensive glossary of continuous improvement terms
- A listing of pertinent websites that specifically address continuous improvement

2 Cycle Time Management

2.1 INTRODUCTION

It should come as no surprise that the primary function of a residential construction company is to build homes. The objective is to maximize the financial performance of the company through the building and sales processes. This can be partially accomplished through several aspects of the construction process, but a few of these can have as great an impact on overall financial performance as the proper management of construction cycle time.

Three major components of cycle time management have a significant impact on the overall performance of the organization: (1) cycle time duration, (2) cycle time variation, and (3) resource management. Cycle time duration deals with how much time it takes to build a home from start to finish, while cycle time variation is the measure of build-time consistency. Both have dramatic effects on home builder resource management and cash flow.

Cycle time management is the acquisition and application of detailed knowledge of how long it takes to build a house (the duration) and how long each house takes when construction of the same or similar model occurs over and over again (the variation from house to house).

2.2 CYCLE TIME DURATION

How long does it take? This is such a common question in everyday life that we frequently forget how essential the answer to the question is in building a house or, more importantly, a community or subdivision of houses. The concept of cycle time duration is applied in both a broad and a narrow sense. It is broad in the sense that it covers the entire process of building a house from horizontal construction at the start of the process to vertical construction at the completion of the process and narrow in the sense that it also focuses on every step, major and minor, in the process.

Thus, a home builder needs to know how long it takes to grade a construction site as well as how long it takes to place a foundation, add in the plumbing, erect the framing, rough in the electrical, all the way to the final steps when the siding or stucco is placed or the final coat of paint is applied.

Why, you may ask, do we place such importance on cycle time duration that it deserves an entire section within this chapter on cycle time management? Simply put, time is money. It is no secret that residential construction is a very capital-intensive industry. It generally requires large sums of money to be spent months or even years prior to starting construction of the first home. On the short end of the spectrum is purchasing finished lots. Vertical construction can begin on these lots as soon as they are purchased because all offsite development has been completed. On the other end of the spectrum is the purchase of raw land. In this scenario, the land must be entitled, planned, and improved before vertical construction can begin.

In either scenario, a significant amount of money is spent prior to starting the first home in any given project. What does that have to do with cycle time management? When you break down the costs, it is easy to see how cycle time reduction can have a large impact on profitability. From a high level, a builder pays for the following prior to receiving payment from the customer: land, improvements, materials, labor, interest, and inflation. Once production is under way, little can be done to reduce the cost of the land, improvements, materials, or labor. However, the amount of interest paid to finance the project and the cost of inflation can be directly influenced by how long it takes a builder to deliver a home to the buyer. Even more importantly, the sooner the home is delivered to the customer, the sooner the cash invested in the home can be recovered and become available to cover operating expenses. Over hundreds or thousands of homes, the financial implications of reducing cycle time can have a major impact on the financial performance of the organization.

2.3 CYCLE TIME VARIATION

How long does it take you to drive to work each morning? Everyone knows that it varies from day to day. Some days take more time while others take less. This occurs as a result of the presence or absence of one or more relevant variables on any given day. It's not a big stretch of your imagination to realize that this thinking also applies to building houses.

When the weather is extremely wet, windy, hot, or cold, the time needed to construct homes is increased. Alternatively, when these same factors occur at moderate levels, then the time required to construct the same homes is reduced. Of course, weather conditions are not the only variables that impact cycle time; we also have to consider variation in trade contractor crew sizes, availability of building materials, as well as availability of both onsite equipment and vehicles. The latter includes scaffolding, delivery trucks, and forklifts.

The big question becomes why is it important to minimize variation? Excessive variation in any process has been linked to lower quality, lower customer satisfaction, and higher costs in many different industries.

From a customer satisfaction standpoint, buying a new home has major life ramifications. It generally requires buyers to move into the new home when completed. Buyers would like to minimize any "double payment" on living expenses so they need to be able to plan when they can vacate their current residence and move into their new home. Excessive variation in cycle time can cause major customer satisfaction problems in this regard. If the variation in cycle time for the given home is plus or minus 10 days, the buyer may end up paying for two places to live for up to 10 days or have no place to live for up to 10 days because they were told their home would close and it took 10 days longer than predicted.

Excessive variation has also been linked to lower quality and higher costs. Excessive variation in cycle time leads to inefficient usage of resources and higher costs. If a framing crew is scheduled to be at a home site on a certain day but the concrete work was not completed in the allotted time, the builder may get additional trip charges from contractors that needed to make additional trips. Conversely, if the concrete was done much earlier than predicted, the home is then just sitting on the construction site and is incurring carrying costs while nothing is happening because the framing crew is not scheduled. What tends to happen is resources get juggled and pulled from one place that is not ready to go to another that is ready but was not necessarily supposed to be. This increases the likelihood of mistakes being made through rushing or quick changes. Instead of having an orderly process where everyone's things are prepared and ready, it becomes a chaotic juggling act that becomes less efficient in the long run. The increased number of mistakes leads to a greater number of defects being passed on to the customer at the time of delivery as well as longer cycle times. They also require more time, money, and resources to correct the problems.

2.4 LEAN CONSTRUCTION

Many industries have realized the importance of cycle time management over the Past few decades. This has led to the development of systems to improve both cycle time variation and duration. One such philosophy or system is lean construction. While the scope of lean construction extends beyond cycle time reduction, it does incorporate many of the key principles and processes necessary to reduce variation and cycle time.

Lean construction focuses on the elimination of all types of waste in a process. Some types of waste are obvious, while other types are hidden and need to be uncovered through analysis. This ranges from wasted materials and wasted motion to wasted resources. Processes that are full of unnecessary (nonvalue-added) steps or procedures and wasted resources are unpredictable and costly.

The elimination of waste in the construction process results in a more predictable process with a shorter duration. To identify the waste in a particular process, you must first know what the current process actually entails. This requires documentation of the current process as it is actually happening. Figure 2.1 displays a common form of process documentation called a process flowchart.

A process flowchart is a visual representation of any process and its individual steps. It is important to fully document the existing "as-is" process as it is actually occurring before significant improvements can be made. To assure an accurate representation of the process, the authors recommend beginning the documentation process by starting at the end and then working backward to the beginning of the process. Doing this forces the mind to think through all of the steps that are actually occurring in more detail minimizing the risk of skipping steps.

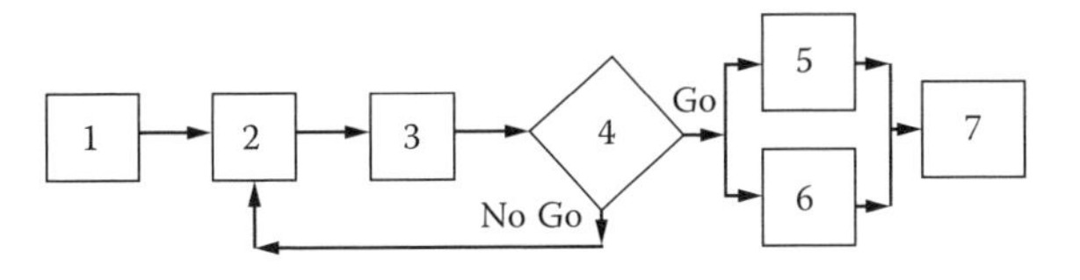

FIGURE 2.1 Process flowchart. Note: 1–7: process steps or tasks.

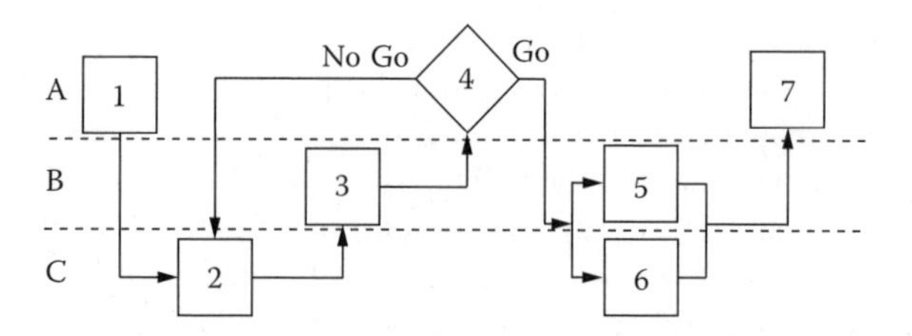

FIGURE 2.2 Process map. Note: 1–7 = Process Steps or Tasks. A–C = Process Stakeholders (Petersons or Departments).

After the process has been fully documented, the process flowchart can be converted into a process map as shown in Figure 2.2. The only difference between a process flowchart and a process map is that the process map illustrates the handoffs between different individuals or organizations throughout the process. This helps identify key transfers of information or work responsibilities between the individuals or organizations.

After creating the "as-is" process map, the next step is to create the process as it "should be" to minimize waste and inefficiency. Each step in the process should be evaluated for its impact on and contribution to the process; that is, is it value added, nonvalue added, or value enabling. Some steps will be indentified as being critical to the process, while others may need to be redesigned and still others should be eliminated. Each step in the process can be categorized into one of the following: value added, value enabling, or nonvalue added. For a step to be categorized as value added, it must directly contribute value to the consumer. An example of a value-added step would be painting. Painting a house directly adds value for a consumer. Thus, eliminating this step would reduce the product's value in a consumer's mind. A value-enabling step is one that does not directly contribute value to the product or consumer but is necessary to continue to another value-added step. For example, installing cabinets would be a value-added step, but it cannot be done without first ordering the cabinets. Ordering the cabinets does not directly provide value to the consumer because the consumer does not care if they are ordered, or stored, or even manufactured internally as long as they are installed in the home. Therefore, ordering cabinets is a step that enables the value-added step of installing the cabinets to occur. Nonvalue-added steps are those that do not meet the criteria of the other categories. These steps are defined as wasteful and unnecessary because they are not directly contributing any value to the consumer, nor are they necessary for any future value-added steps to be completed.

After categorizing each step, the nonvalue-added steps should either be eliminated or structured so their task durations are minimized. Through elimination of wasteful steps, their contribution to both the cycle time duration and variation will be eliminated from the overall cycle time. However, this is not the end of task. There may be other opportunities to improve the remaining value-added steps.

Creating the process as it should be allows you to integrate new technologies that have become available and that may provide the same value to the consumer but will reduce the task duration compared with previous methods. Using the painting example mentioned earlier as a value-added step, it did not get eliminated with the nonvalue-adding steps, but it does have task duration of "x" days. One method for painting the house would be to use a paintbrush and a bucket of paint. Many years ago, that would have been the most modern method of painting. However, over time, spray paint technologies have been developed to greatly reduce both the number of people and the amount of time needed to paint a house. The impact of incorporating the new technologies or methods can significantly reduce the duration of the task. Combine the duration savings from three or four tasks, and the overall reduction can be substantial.

While creating the should-be process, the as-is process should be evaluated to identify tasks that can be performed at the same time (simultaneously) rather than one after the other (sequentially). The set of tasks that must be performed in a particular sequence is referred to as the *critical path* in project management terminology. The critical path duration dictates the overall cycle time of the project. Tasks that are necessary, but not part of the critical path, can be accomplished in parallel with other tasks (simultaneously or at the same time) without adding to the overall cycle time duration. Through identification of the critical path (or sequence of tasks that must be performed in a specific order) and the performance of all other tasks in parallel, some bottlenecks that previously slowed down the construction process of the home will no longer present as great an issue because they are not impeding the progress of other tasks.

However, bottlenecks on the critical path will be of supreme importance to manage and deal with because they will directly increase the overall cycle time duration. Gantt charts can be used to track and manage the sequences and durations of tasks in a project. Many project management software packages are available to facilitate this process. This topic is discussed in Chapter 12 on software.

Mistakes and rework that occur throughout the process also contribute to cycle time variation and duration. After eliminating unnecessary tasks, consider arranging the remaining tasks in a way that reduces the number of mistakes or the amount of rework necessary to minimize variation. The Japanese refer to this process as *Poka-yoke,* or mistake proofing. This can be achieved at any level of the task, e.g., the arrangement of tools in a truck so as to reduce the likelihood of grabbing the wrong tool or having to go back to the original process order when information is missing.

It is important to keep in mind that certain mistakes or errors can be avoided if the process is structured to prevent them from happening in the first place. An exaggerated example would be not waiting for the framing inspection certification prior to installing the drywall. While it is unlikely that any construction company is making this mistake, other tasks may be arranged in a similar fashion, causing significant rework or mistakes to be made.

After going through the illustrated exercises, the "as-is" process should be ready to finalize and document. It is always advisable to conduct a pilot test of the new process prior to full-scale implementation to ensure the proposed improvements actually lead to real improvement, as well as to catch any unforeseen consequences of the adjustments. Once the pilot test is complete, the new process is ready for full implementation.

This process can also be used on an individual task level. The entire construction process is composed of many smaller processes coming together. Many of the tasks performed at the construction site are only the final phase of a larger process for each subcontractor. Concrete shows up at the home site ready to be poured, but many other steps are required to make that happen. Each of these processes can be evaluated individually using the same techniques and procedures illustrated in this chapter. The net effect will be a shorter cycle time from beginning to end that is more predictable and performs at a higher level of quality production. The time, personnel, and cost savings that result from cycle time management will be well worth the investment.

2.5 EXAMPLE

Home builder A is experiencing a surge of cancelations at a number of projects due to the inability to complete homes in a time frame acceptable to the customer. The surge in cancelations has resulted in an excess of inventory homes that will have to be reduced in price to be sold. The lost revenue and carrying costs are taking a toll on the organization and cannot continue. Management is unsure if the extended cycle times are a result of poor employee or contactor performance, or a more systemic problem in the schedule. A process improvement team (PIT) has been organized to understand and fix the problem. At the first meeting of the PIT, before team members attempt to fix the problem, they recognize they must first fully understand the current situation. Thus, they decide to document the construction process.

On their first attempt to document the process, they started at the beginning of the process and sequentially worked through the construction process. When they reached the final step, the ensuing discussion about the tasks in the process made it clear to all that they had not documented the entire process and had skipped many small but important details. They revised their approach and began their documentation of the "as-is" (current) process, working backward from the delivery of the home to the very first step. After two backward passes through the construction process, the team was confident that it had accounted all of the relevant details.

The PIT then began to evaluate each task as it was currently being performed to identify steps or tasks that could be eliminated without affecting the quality of the completed homes. After completing the evaluation, the team determined that there were a few unnecessary steps but not enough to explain the extensive delays being experienced. Next, team members evaluated the actual durations of various scheduled tasks and subsequently determined that the problem seemed to be a recently implemented procurement/purchasing system that had been implemented to reduce material prices. The new system, while reducing the cost of materials, failed to ensure delivery of the materials to home sites on schedule. The right people with the right skills and equipment were ready to perform the work, but the materials were not available and thus were delaying the entire construction process. With some simple adjustments to the new purchasing process, significant cycle time reductions were realized, both for the builder and its trade contractors.

3 Problems

3.1 INTRODUCTION

Every organization, no matter what its size or industry, faces challenges on a daily basis. Some challenges are significantly greater than others. Some seem to repeat themselves, while others seem to come out of the blue when you least expect them. No matter how big or small, they all have a cost. Be it in increased man-hours due to rework, wasted materials, paperwork errors, miscalculations of revenue, or lost customer satisfaction, all problems impact bottom-line profitability to some degree.

The reality is you can't prevent all problems from ever occurring, but through the timely application of the most appropriate process improvement techniques, the frequency and impact of common problems that account for most of the wasted time, energy, and ultimate cost of inefficiency in the residential construction industry can be minimized. It may seem almost counterintuitive, but solving the problem at hand in many cases is the easiest part of the entire process; the difficult part is identifying which problem to solve and which process improvement tool, technique, or method to use.

The following chapters address the issue of limited resources and provide tools and guidelines for solving problems, but this chapter focuses on exploring the different types of problems you may find yourself facing in the residential construction industry and how to decide which problem is the highest priority.

3.2 IDENTIFICATION

To identify a problem we first have to define what a problem is. In the context of this book, problems are situations that result when processes consistently generate suboptimal results. Or, to put it more plainly, the actual results are consistently different from the desired or expected results.

Problems occur in every aspect of the business, and sometimes we feel as though they are virtually never ending. As soon as one problem is solved, there are new problems to take its place. This section briefly covers some of the most common problems faced within the residential construction industry. This is not a comprehensive list, but rather it is intended to start you thinking about how and where the tools and skills covered in this book can be applied to achieve results.

No matter what the problem is, there are two keys to identifying problems. First, you must be able to recognize when a result is outside the range of desired outcomes. This can be done in a number of ways depending on the situation. This is generally done by monitoring various business performance metrics. The second aspect of problem identification is assessing the impact it has on the business. As the old axiom asks, "If a tree falls in the woods and no one is around to hear it, does it make a sound?" a problem should have some kind of adverse effect on the business itself. If you cannot define how or why it has an effect, you have to ask yourself, "Is this really a problem?" In this case you may still have a problem, but it is a result of using the wrong metric or identification tool.

In illustrating this concept, let's start with what was a hot litigation topic in the early 2000s: construction defects. For the purpose of this book, a construction defect is some aspect of a home that does not meet engineering specifications or customer expectations. This can happen at any point throughout the construction process. On the construction site, the field manager's role is to identify problems associated with the homes for which he or she is responsible. These problems are generally identified through inspection of the home or the work being performed. On an individual level you know there is a problem with that particular home being built, but you don't know if the problem is more widespread and systemic in nature.

Management, on the other hand, is more interested in identifying and resolving problems that are more pervasive or occur with excessive frequency. To identify widespread problems, management must use performance metrics that summarize the entire portfolio instead of just individual cases. One commonly used metric is frequency of occurrence. For the field manager, one occurrence is enough to warrant attention and remedy. Alternatively, management is interested in identifying problems that are occurring across all homes under construction and that require a change in the process of construction to reduce the likelihood that the problem will recur in the future. Table 3.1 provides a list of a few common problems as well as potential metrics that are used to identify those problems.

3.3 RANKING

Once a problem has been identified, how do you decide how important it is, especially with respect to other potential areas of concern? If you have only one problem to deal with, then by default, it is the top priority. However, when you have multiple problems to deal with, being able to rank them in order of importance (using any

TABLE 3.1
Problems and Methods

Potential Problem	Potential Identification Metrics
Rework due to home damage	Frequency, cost, cycle time delay
Missed schedules	Frequency, cycle time delay
Theft	Frequency, cost
Safety violations	Frequency, severity
Purchase orders for work	Frequency, cost
Cycle time	Average cycle time, variation in cycle time
Customer satisfaction	Average satisfaction score, percent satisfied customers

one of a number of potential criteria, such as cost, time, customer satisfaction level) will allow you to make the greatest positive impact on the business in the shortest amount of time.

Cost is a commonly used attribute for ranking or prioritizing problems. This metric is easily understood and does a good job of identifying which problem is hurting the business the most right now, but it is not without its drawbacks. Ranking problems solely on their current financial costs does not account for strategic or future implications of other problems. Let's take problems such as concrete repair and low customer satisfaction as examples. It's easy to quantify how much it is costing the business to fix slabs, driveways, or sidewalks. It is more difficult to quantify how much it is costing the business through resulting low customer satisfaction levels. In theory, higher customer satisfaction should lead to higher profitability in the future through repeat customers and referrals, but as of yet that link has been difficult to quantify. Even if it were fully quantified, low customer satisfaction scores may or may not impact the business immediately; it may take months or years to see the impact. Using only the cost metric in prioritization of problems leaves your company vulnerable to ignoring the larger long-term and strategic problems that are harder to quantify but that could have a far greater impact on the overall health of the company.

Multivoting is another quick and easy way for a group or team (there really is a difference) to determine the highest priority items within a defined list. Its strength is its simplicity, and it is best suited for prioritizing long lists or when the ranking is being done by a large number of people. In this technique, each person can vote X times, where the value of X is approximately one third to one half of the total items that need to be prioritized. Team members then vote individually for the items they believe to be of highest priority or of greatest importance. Eventually, the list is culled down to the top four to six items for further discussion and ranking against each other. While you can continue to use multivoting to determine your final rankings, you may wish to consider using a more quantitative or criterion-based ranking method for prioritizing the top items, such as paired comparisons or a prioritization matrix.

Paired comparisons (also known as pairwise ranking) is a structured method for ranking a small list of items in priority order. This technique is similar to determining a regular season division winner in professional sports. Each item is pitted against every other item individually, and the team decides which is more important (or the winner). After every item is compared with every other, the item with the most "wins" would be ranked first, the item with the second most wins would be ranked second, and so forth. This method is more complex than multivoting, but it does reduce subjectivity and provide more structure. It is recommended for use when there are a small number of items to prioritize.

Another approach would be to use a prioritization matrix. In this approach, the team develops a list of criteria by which each item will be evaluated. After the team agrees on the list of criteria, it then decides how each of the criteria should be weighted. The items are then evaluated and scored using the weighted criterion with the highest score being ranked first in importance and so on. This method is the most objective ranking procedure because each item is evaluated using consistent, pre-identified criterion, but it is also the most complex. Detailed, step-by-step descriptions of the previously discussed techniques are provided in Chapter 13.

TABLE 3.2
Situation vs. Ranking Methods

Method	Conditions
Multivoting	Many items to rank
Large group of participants	
Only need to separate the top group of items from the rest	
Some subjectivity is tolerable	
Need to prioritize quickly	
Typically a precursor to paired comparisons or prioritization matrix	
Paired comparisons	Small group of items to prioritize
Absolute rank order is important	
Need to prioritize quickly	
Some subjectivity is tolerable	
Prioritization matrix	Small to moderate number of items to prioritize
Absolute rank order is important	
Strict objectivity is required	
Capable of spending the time to perform thorough analysis	

Table 3.2 provides a quick reference guide for deciding which ranking method to use based on the situational demands.

3.4 EXAMPLE

Let's simulate (only kids pretend) you are responsible for leading improvements in customer satisfaction for your organization. The company has historically performed slightly better than the industry average, but there is considerable room for improvement. Senior management announces a new focus on improving customer satisfaction quickly. You proceed to assemble a cross-functional team of approximately 10 to 15 members and develop a substantial list of problems that need to be addressed to improve customer satisfaction. In this situation, there is a relatively large team with a long list of items. The team has the capacity to deal with several items simultaneously, so the objective is to separate the top four or five items from the remainder. Absolute rank order is not as important in this situation because all of the top items can be addressed. Multivoting would be the most effective method to apply in this example.

After narrowing down the list, you break the team into subgroups based on functional areas, and the items are then distributed to the appropriate teams. One team ends up with four items to address but has the capacity to effectively address only two of the four items at one time. The subgroup is now a much smaller team and must rank the four items in terms of relative importance given the objective of quickly improving customer satisfaction. The group then goes through the paired comparison exercise to identify the top two items to address in their functional area.

4 Problem Solving

4.1 INTRODUCTION

The ability and desire to solve problems is an innate trait found in the human species. One could argue that some people are better than others when it comes to problem solving, but we all seem to share a raw ability and drive to overcome the challenges we face. It goes without saying that the problems we face today are far more complex than those faced in ancient times or even 50 years ago, but it is interesting to think about the rate of advancement in civilization over the length of modern history. For centuries and perhaps even millennia, the primary drives of food and shelter dominated life's daily problems. At first the problem of obtaining food was addressed by using bare hands; then came the sharpened stick. This was followed by the stone-tipped spear, then the bow and arrow, and finally modern firearms. Such improvements throughout the course of history were not random; however, the rate of technological progression seems to have increased exponentially in the modern day. A hundred years ago we had motor cars that could go 30 miles per hour; today we can launch a rocket into space and hit a comet 267 million miles away traveling at 64,000 miles per hour (NASA Deep Impact Mission) with the comet traveling at high speed in the opposite direction. How has this happened?

Albert Einstein proclaimed that we can't solve problems with the same level of thinking that was used to create the problem. Does this mean people as a species are becoming genetically smarter than the previous generations given the rapid advances in technology? While it may be an appealing thought to some, it is probably not the case. It is more likely that the difference between today and even 100 years ago is not due to genetically smarter people; it is because of the way we approach problems. Over the past 100 years, the tools, techniques, and methods for problem solving have become much more scientific in their approach. This does not mean that mistakes will not happen; it means that methods have become more systematic and controlled. This systematic learning and problem-solving approach allows you to build on a body of knowledge instead of reinventing everything. It also allows you to identify and address the true or root cause of the problem instead of just the symptoms. This concept and technique is discussed in detail in Chapter 6, "Root Causes."

This chapter introduces five of the most common problem-solving methodologies used today. You will probably notice that fundamentally they all share the same core philosophy; they just vary in the details and application.

4.2 OBJECTIVES

The problem-solving methodologies introduced in this book can be categorized into two groups based on their objectives. The first group focuses on reducing the variation and defects in a process or product, and the second group focuses on eliminating waste. Before introducing the methodologies themselves, it is important to understand why reducing variation, defects, and waste are important as well as which should be addressed first.

4.2.1 Accuracy versus Precision

If you were given a choice of managing one of two different processes where one process was accurate but not precise and the other was precise but not accurate, which would you choose? You might ask what is the difference, or what on earth are you talking about?

To keep things simple for the time being, let's say you are playing darts and you need to choose one of four people to be on your team. You ask individuals to throw a few practice rounds to evaluate their performance, asking them to aim for the bull's-eye. The comparison of practice throws among Persons A, B, C, and D looked like the four results shown in Figure 4.1.

- Person A's darts are both unpredictable (large spread) and inaccurate (off target). There is too much variation in A's throws. Sometimes it hits the target, while other times it comes close but misses.
- Person B would be described as accurate (on target) but not precise (large spread) because the darts landed in the general area of the targeted bull's-eye but are too spread out.
- Person C would be described as precise (small spread) but not accurate (off target) because his or her throws were grouped very close together but not sufficiently close to the targeted bull's-eye.
- Person D, on the other hand, is quite accurate (on target), and his or her throws are close to the target (small spread). Thus, there is a real confidence that his or her throws will end up centered and clustered around the bull's-eye.

Of course the person to select is D, but just suppose there was no Person D and a choice needed to be made; which of the three remaining persons would be the "best" choice?

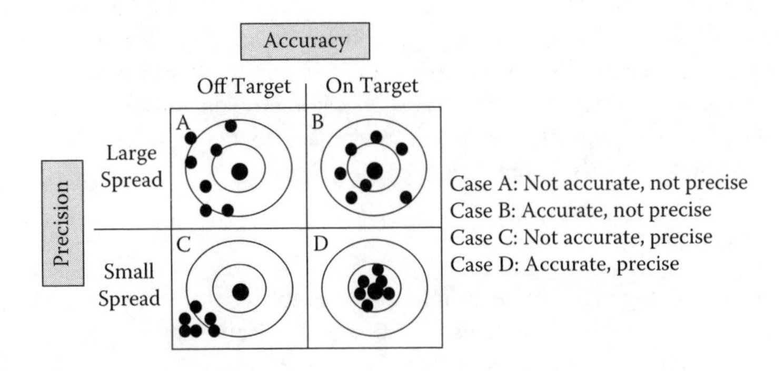

FIGURE 4.1 Accuracy versus precision.

In this situation, we would recommend selecting Person C. This may surprise some readers, but there is a good reason. The key is reliability; Person C is quite reliable because there is such little variation in the outcome of his or her throws. Therefore, when an adjustment is recommended to his or her throwing process such as a change in aim, he or she will be throwing bull's-eyes all day long. Alternatively, Persons A and B are much less reliable, and, therefore, when an adjustment is made to either of his or her throwing processes, it would be much more difficult to assess whether the adjustment made things better or worse.

A predictable process with little variation is much easier to improve than an unpredictable process. Therefore, when it comes to process improvement and problem solving, reducing variation to increase predictability and then working to improve accuracy is far more effective in delivering results than just trying to improve accuracy. Ignoring variation inevitably leads to a trap of continuous change but not necessarily to improvement. When the variation masks the effect of the change, people are making decisions based on random variation instead of predictable output, which leads to seemingly endless changes without visible improvement.

The question then becomes what should your organization focus on first, that is, right now? We have developed a brief checklist in Figure 4.2 to help you determine whether it is more beneficial for your residential construction company to focus on reducing defects and variation through the judicious application of Six Sigma or to focus on reducing waste to increase profitability through the use of Lean.

Tool	Question	Poor			Average			Excellent
6σ#1	Rate your overall product quality compared to other builders	1	2	3	4	5	6	7
6σ#2	Rate your level of actual warranty costs	1	2	3	4	5	6	7
6σ#3	Evaluate your amount of rework	1	2	3	4	5	6	7
6σ#4	Number of incomplete items recorded at delivery of home to customer	1	2	3	4	5	6	7
6σ#5	Ability to consistently meet commitments to customers	1	2	3	4	5	6	7
Lean #1	Overall profitability	1	2	3	4	5	6	7
Lean #2	Overall constructions costs	1	2	3	4	5	6	7
Lean #3	Level of standardization across company	1	2	3	4	5	6	7
Lean #4	Correct utilization of talent within the organization	1	2	3	4	5	6	7
Lean #5	Level of bureaucracy or red tape in organization	1	2	3	4	5	6	7

FIGURE 4.2 Six Sigma–Lean comparison checklist.

To maximize use of the checklist, evaluate your organization on the 10 aspects noted in Figure 4.2. Each aspect will be rated using a scale of 1 to 7 with 1 being poor, 4 being average, and 7 being excellent. Five aspects are associated with Six Sigma, and five are associated with Lean. After evaluating each aspect, sum the scores for Six Sigma and Lean separately. Since the rating scale uses higher numbers to indicate better performance, the lower score will indicate which should be addressed first. For example, if an organization scored a 29 on the Six Sigma questions and a 17 on the Lean questions, it would most likely be more beneficial to the organization to implement a Lean strategy with its limited resources before embarking on Six Sigma implementation.

4.3 METHODOLOGIES

4.3.1 Total Quality Management (TQM)

Since its beginning in the early 1950s, total quality management (TQM) has become a widely practiced and widely misunderstood approach to continuous improvement. One of the foundations of TQM is the problem-solving method: plan, do, check, act (PDCA) cycle (Figure 4.3), PDCA is also sometimes known as the Deming cycle or the Shewhart cycle.

PDCA is an iterative and systematic methodology for problem solving. PDCA iterative in nature because once the last step has been completed, it loops back to the beginning and starts over. It is also a very systematic method of problem solving because it is based on the scientific method of forming a hypothesis, performing an experiment, and then evaluating the results to draw conclusions. These two aspects are what make PDCA so successful. This methodology allows you to build upon prior knowledge to start the next and more complex cycle of incremental improvement.

- Plan: The planning phase consists of designing a change or inspection to a product or process with the intention of improving the product or process.

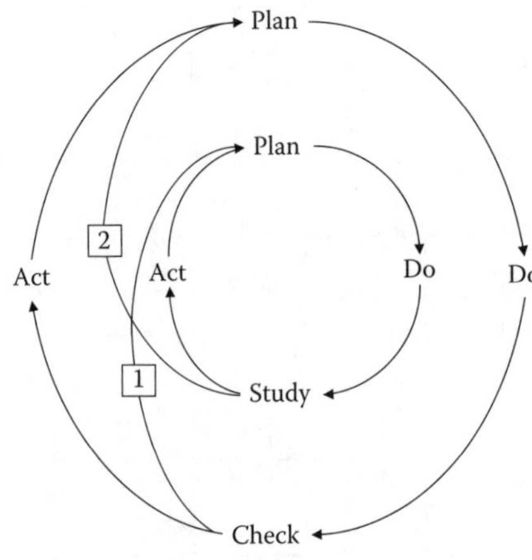

- Plan-Do-Check-Act (PDCA) [the outer cycle] is used to **standardize** a process (for consistency/uniformity).
- Plan-Do-Study-Act (PDSA) [the inner cycle] is used to **improve** a process (either Kaizen or Process Re-design).
- The "orbital leaps" should take place when it is necessary to move from a standardized process (PDCA) to find an improved process (PDSA) [Leap 1] or having created an improved method, to standardize it [Leap 2].

FIGURE 4.3 PDCA cycle.

In this phase you will analyze the process or product to be improved by identifying problems or potential areas of improvement. Common diagnostic tools used in this phase (discussed in detail in Chapter 13) include flowcharts, process maps, and Pareto charts.

- Do: This phase consists of implementing the planned change to the process or product. Oftentimes it is preferable to implement on a small scale first until you have completed the entire PDCA cycle and have determined the effect of the changes made. Not surprisingly, not all changes result in positive results. Testing on a small scale limits the organizational disruption of this process by implementing on a large scale only changes that produce quantifiable positive results.
- Check: This phase consists of analyzing the results of the change or test. Did the planned change yield results? Were the results positive or negative? To answer these questions, you must rely on predetermined metrics designed to measure the effect of the change. Some of the possible tools that could be used to detect changes are run charts, control charts, and box and whisker charts.
- Act: In this phase you decide either to adopt the change and implement it on a broader scale, to abandon the change, or to run it through another cycle.

4.3.2 Six Sigma

Six Sigma is another scientifically based problem-solving methodology that has gained tremendous popularity and results over the past 15 years. Developed by Motorola in the early 1980s and adopted by companies such as General Electric (GE) and Honeywell in the early 1990s, Six Sigma has become a part of major corporations around the world.

The Six Sigma methodology shares many of the core characteristics of the other problem-solving methodologies. What makes it different is the strict reliance on data in decision-making processes, the focus on quantifying the financial impact of Six Sigma projects, and the emphasis on meeting specifications and expectations as defined by the customer. The Six Sigma problem-solving methodology consists of the following five phases: define, measure, analyze, improve, and control (DMAIC).

4.3.2.1 Define

The first step in solving a problem is properly defining the problem. If you have a problem with your trench to close of escrow cycle time, exactly what are you trying to solve? Is the duration too long? Is it too unpredictable or inconsistent? To effectively solve the problem, it has to be clearly defined in specific terms. There are several inherent benefits in explicitly defining a problem. First, it helps to make sure everyone on the team is working on or toward the same goal. Second, it helps prevent teams from overcomplicating the problem-solving process by trying to solve three or four different problem at once. Consider the following problem statements:

1. Improve the cycle time.
2. Reduce the average cycle time from layout to frame inspection by 10%.
3. Reduce the variation in cycle time from layout to close of escrow by 10%.

Both statements 2 and 3 could be included in statement 1. If you defined your goal as only "improving cycle time," where would you start, and how would you know if you succeeded? Clearly defined problems are essential for sustained continuous improvement.

Being specific in your definition is necessary, but it also has to be relevant to the customer requirements or expectations. In the define phase you gather and define the critical internal or external customer needs and requirements relevant to your problem. In Six Sigma terminology it is often referred to as the *voice of the customer.* While the importance of listening to the customer seems somewhat obvious, it is startling how often this vital aspect is skipped or ignored in the heat of the battle of problem solving. There is a tendency to try and figure out the optimal solution that would make things easiest for the organization to deal with instead of optimizing what would result in the greatest value to the customer. It is important to remember that the purpose of a business or organization is to maximize shareholder or stakeholder value. You can't maximize shareholder or stakeholder value by ignoring or alienating customers and their needs.

4.3.2.2 Measure

Once you have clearly defined the problem to be solved, the next step is to determine how the process is performing in its current state. This will become your benchmark for improvement. Using the previous examples, for problem statement 2 we would measure the current average days between layout and frame inspection. Once we've calculated the average cycle time, we have the baseline performance metric as well as the target improvement metric. For problem statement 3, instead of focusing on the average days between events the focus will be on the standard deviation. It is important to choose the proper metric in both the define and measure stages. Dealing with data and metrics are discussed in detail in Chapter 5.

4.3.2.3 Analyze

At this point, you have a clear understanding of what needs to be accomplished by defining your objective. You know how the process is performing in its current state through measurement. Now the process should be analyzed for potential areas of improvement that would directly contribute to reaching the defined objective. Frequently used tools in this phase include Pareto charts, process flowcharts or process maps, scatter analysis, and defect maps. This step focuses on defining the cause and effect relationships within the process and making sure that all possible variables have been considered and assessed.

4.3.2.4 Improve

After analyzing the process the next step is to develop and implement planned improvements to the process based on the data collected by your team. The improvements may be based on data collected in the analyze phase such as identifying elements of a process where delays are frequently occurring. It could also be done through design of experiments, where a series of tests involving multiple variables is used to determine the ideal situation and results.

4.3.2.5 Control

Once the improvements have been made, the final step is to evaluate the results and to monitor the process for defects or deviations from the desired results. This can be done through control charts or other tracking metrics.

4.3.3 Lean Six Sigma

Lean Six Sigma is the blending of lean manufacturing and Six Sigma. While traditional Six Sigma programs focus on eliminating variation and reducing defects, lean Six Sigma focuses on the efficient use of resources in the production process through the elimination of waste. By eliminating waste the efficiency of the production process is increased, which increases production capability and quality while reducing costs.

In the residential construction industry, it is easy to think of waste in terms of materials and time, but lean Six Sigma goes far beyond that typical concept or definition of waste. In this paradigm, waste is defined as any activity that uses resources of any kind but does not provide value to the customer. The types of waste can be classified into two different primary categories: obvious waste and hidden waste.

Obvious waste is easily identifiable and recognizable. These are typically things that can be eliminated relatively quickly and at a minimal cost. Hidden waste, on the other hand, refers to aspects of a process that appear to be necessary or contributing value under the current process but that could be eliminated through process redesign or improvements in technology. Both obvious and hidden waste can be further broken down into the following categories:

- Wasted human talent/skills: This type of waste happens when people in the organization are used in a capacity that does not match their strengths. This does not necessarily mean an individual is performing at a less than an adequate level but simply that he or she has talents and skills that are more beneficial to the organization and that are not being used. An exaggerated example of this would be having an employee trained in financial analysis working as a model home maintenance worker. While the employee may be quite capable in his or her current position, his or her more valuable skills are not being used and therefore are being wasted. Another situation occurs when an employee possesses a talent or skill; the organization does not realize the true value of that skill, and therefore that capability is underused.
- Defects and mistakes: Correcting defects or mistakes adds unnecessary costs through the requirement of additional materials, equipment, and labor. Other potential costs due to correcting problems include delays in schedule that could pull resources from production of other homes and could incur unnecessary incremental interest charges. Defects and mistakes can also have an impact on customer confidence and satisfaction. Erosion of consumer confidence and low customer satisfaction inevitably lead to a reduction in sales and profitability. If you can't satisfy your customers, sooner or later someone else will.

- Excess inventory: Inventory in this sense is defined as either finished products, works in progress (WIP), or raw materials. The relevance of this type of waste in residential construction depends on the structure of the organization. For builders who use subcontractors to provide and manufacture the components and materials used in the home, there is not much raw material storage going on. For builders who have integrated aspects of the homebuilding into their organizations, this can be a greater issue. Materials such as concrete aggregate, lumber, and electrical wiring all take up space and require monitoring for theft prevention. They also must be moved from location to location multiple times, which consumes resources and equipment but provides no additional value to the finished product.
- Overproduction: In the residential construction industry, overproduction is a form of waste that can cripple or even bankrupt an organization. Residential construction is a capital-intensive industry. It requires large amounts of cash to be spent before any can be recovered. While sale of the finished homes returns substantial amounts of cash, there is a great deal of pressure to recover that cash as quickly as possible. This can lead to overproduction of speculative units that do not have buyers. In healthy market conditions, speculative units are a financially beneficial method of product segmentation and revenue generation. It meets the needs of buyers who cannot or are not willing to wait for a home to be built from dirt, and it allows the company to maintain construction resources in its community. However, overproduction in it mildest form results in reduced margins by requiring discounting of the product to turn it into cash and to recover the money invested to build it and prevent further interest payments. In its most severe form, as recent history has shown, overproduction can result in massive price declines for complete markets leaving entire organizations struggling to generate the cash necessary to continue operations because so much of the value of the organization is tied up in unsold homes under construction or unsold speculative units.
- Waiting time: Time spent waiting without productive activity is something that everyone has experienced but probably does not realize how much waste is actually occurring. Waiting time is a waste because valuable resources are sitting idle while anticipating the arrival of information or another part of the process to be completed. This can happen in any aspect of the business. In administration it could be waiting for paperwork from a co-worker or paying a second trip charge to the painter because the drywaller was not finished. Frequent causes of wasted time waiting are inefficient process, poor quality upstream in the production, uneven or inaccurate scheduling, unreliable subcontractors, or even poor equipment.
- Unnecessary motion: Any unnecessary motion of equipment or people that does not produce value to the product or service is a waste. Unnecessary motion in equipment causes wear and tear, and unnecessary motions from people can cause fatigue and increase the likelihood of injuries. This is often evident through poor human–machine interface designs or workspace design.

- Unnecessary transportation: Any unnecessary transportation of materials, people, equipment, or information that does not directly support actions that add value to the customer is waste. One of the most personally frustrating examples is e-mail carbon copies to unnecessary people. Such e-mails waste resources and time and disrupt productivity. Other examples include unnecessary back-and-forth from the field to the office. Such trips increase wear and tear on equipment and tie up resources in activities that provide no value to the customer in the end.
- Processing waste: This type of waste occurs when there are unnecessary processing steps or materials. Things like multiple approvals and unnecessary forms or information capturing are wasted activities.

4.3.4 Kaizen

Kaizen is a Japanese concept that, in its fullest concept, applies continuous improvement to all aspects of life: personal, home, social, and work. At work, this encompasses everyone in all aspects of the work environment. Kaizen is sometimes seen as a method for organizing workspaces such as offices or shop floors. While it is most easily adapted to this situation, it should not be viewed as being exclusively a "housekeeping" methodology.

Kaizen focuses on improving efficiency and morale by eliminating wasted time and movement and minimizing errors through the 5 Ss: *Seiri* (sort), *Seiton* (set in order), *Seiso* (shine), *Seiketsu* (standardize), and *Shitsuke* (sustain).

- Sort: The sorting phase is truly the housekeeping phase of the entire process. In this phase all materials, tools, files, and so forth should be evaluated and only essential materials kept. All nonessential materials should be stored elsewhere or discarded.
- Set in order: This phase focuses on creating efficiency in workflow. Tools, processes, and materials should be arranged to promote workflow from one step to the next. For example, tools should be kept where they are going to be used.
- Shine: The shine phase focuses on cleanliness. This should be a part of the daily routine, not a once-in-a-while undertaking. Every day the jobsite should be cleaned and tools put away. This ensures that everything is where it should be when work resumes. There is no time lost due to misplaced equipment. It also helps promote quality through attention to detail and promotes respect for the work and the home being built.
- Standardize: Standardization is used to ensure that all people know their responsibilities and what they should be doing. This is a result of standardized processes, operating procedures, and work practices. Everyone has the same understanding of what should be done as well as what is acceptable and what is not.
- Sustain: Sustaining is the maintaining and reviewing of the standards or the new way of working. The objective is to prevent the workplace from regressing back to old ways through maintained focus on the established

standards and practices. This is not to say that improvements and new suggestions should not be listened to or incorporated; they should just be evaluated and implemented within the structure of the 5 Ss to prevent regressing to a less efficient work environment.

4.2.5 B.U.I.L.D.

We recognize the unique nature of the residential construction industry. To this end we have created a problem-solving methodology especially designed to identify, quantify, and resolve undesirable situations specific to home building. The acronym for this methodology is B.U.I.L.D., which is the first letter in each of the following concepts:

- **Begin with process pain:** When a person has been hurt in an event such as a car crash or a construction accident and is then brought to a hospital emergency room, the first thing that usually happens is for a triage nurse to identify and prioritize all the physical damage that occurred to the person as a result of the accident. It's no different when there are multiple calls by a collection of homeowners, who are all complaining about problems they are experiencing with their new homes. At this point, you—the home builder—need to perform triage to decide which construction process pain requires some of your limited critical resources that must be deployed to address those processes which need corrective actions. Once the various

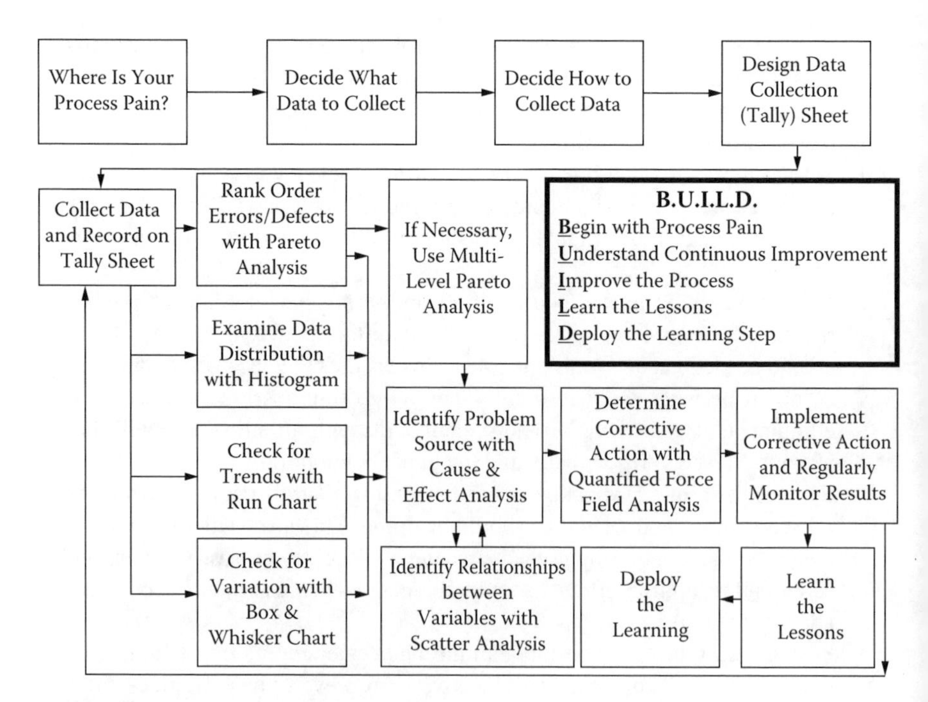

FIGURE 4.4 B.U.I.L.D. problem-solving sequence.

process pains have been identified and prioritized, the next logical step is to introduce continuous improvement.

- **Understand continuous improvement:** As explained earlier in this chapter, continuous improvement (or Kaizen) is really nothing more than deciding to improve a process such that the measurable outputs of the process achieves greater success in the future than before. Understanding continuous improvement requires training, study, and practice to develop the confidence and insights necessary to apply the correct tools at the correct time and in the correct order. We believe it's faster, less expensive, and more practical for a home builder to seek whatever counsel and assistance is necessary from those who already have the experience of applying continuous improvement principles and tools in residential construction.
- **Improve the process:** Using the tools and techniques of continuous improvement as described in Chapter 13, the concerned and responsible home builder establishes a baseline (the "as is" condition) for each critical process so as to have a starting point for the continuous improvement pathway.
- **Learn the lessons:** Once the targeted process has been improved such that the critical performance metrics have achieved at least the original goals, the home builder's team of improvement specialists should document the entire sequence of events from "B" to "D." Prior to moving to the final step, confirm the results by replicating whatever the team has done. If the replication does not provide similar results, then the team is advised to return to the initial step and begin again. If the replication confirms the earlier results, this should be so stated in the team's report. The final document provides the entire company with a historical record of what the specific problems were and how the team went about resolving them.
- **Deploy the learning:** Thus, once the lessons have been learned, the wheel no longer needs to be reinvented. Share the wisdom throughout the company, making sure that the team's documented report is easily accessed and available to other individuals and teams that can put the information to good use.

4.4 EXAMPLE

Builder A is struggling to remain competitive in a highly competitive market. Costs keep rising, and margins keep shrinking. Something must be done to improve the organization's performance. Because available resources to implement global initiatives are somewhat limited, management has decided it is best to implement a focus on either Six Sigma or Lean manufacturing. To decide which would be the most beneficial to implement first, members of management review and evaluate their performance using a number of business metrics.

As they review their customer satisfaction data, they note that the customer ratings for quality are below expectations, but the service scores or ratings received by the employees is good. When reviewing the company's financial performance in detail, it was noted that construction costs were higher than the market average for a comparable product. The major contributors to the increased costs were purchase orders for work outside the scope of master contracts, repairing home damage that occurred during the construction process, and high warranty costs. In this case, personnel and overhead costs are within normally accepted ratios.

The data review led management to conclude that an emphasis on improving product quality as well as reducing rework and errors would have a greater impact on the overall value of the organization than would Lean manufacturing. Six Sigma programs designed around reducing construction-related damage and rework and improving quality to meet customer expectations on a more consistent basis will address many of organization's most pressing needs.

5 Dealing with Data

5.1 INTRODUCTION TO DATA

This information is provided to ensure that the reader is aware that most problem-solving tools and techniques are intended for use with specific types of data. For example, which equation should be used to calculate a minimum sample size depends on whether the data are considered to be attribute (discrete) or variable (continuous). Selection of the correct equation also depends on whether the data are drawn from an infinite or finite population. These topics are discussed and demonstrated in detail in Chapter 11.

5.1.1 WHAT ARE DATA?

Depending on your source of information, *data* are defined in many ways, including the following:

- A collection of facts from which conclusions may be drawn (e.g., statistical data).
- A datum is a statement accepted at face value (i.e., a given). *Data* are the plural of datum. Large classes of practical, important statements are measurements or observations of a variable. Such statements may be composed of numbers, words, or images.
- A representation of facts, concepts, or instructions arranged in a formalized manner suitable for communication, interpretation, or processing by humans or automated means.

5.1.2 TYPES OF DATA

The following provides a discussion regarding ways data have been categorized.

5.1.2.1 Attribute/Discrete Data

Attribute data, also known as discrete data, are generated by counting, for example, the number of windows or doors in a sample of x homes that allowed water intrusion or the number of homes that closed during a specific calendar period.

- An *attribute* is the presence or absence of a particular characteristic. Working with products, services, and processes, things are classified as, for example, good versus bad, accept versus reject, or go versus no-go.
- Also known as *discrete* data, attribute data are generated by counting using whole numbers (i.e., no fractions or decimals). For example, Community

X has 10 two-story homes and 8 one-story homes, and Builder Y uses 33 trade contractors to build a community. Another example would be Builder Z experiencing 12 lost time accidents this year and 19 last year.
- Attribute data are easier and faster to collect and record than variable data; however, they do not provide nearly as much information as variable data.

5.1.2.2 Variable/Continuous Data

Variable data, also known as continuous data, are generated by measuring, for example, the cut length of a stud intended for placement in an interior archway or the number of cubic feet of concrete to be placed in a foundation. The accuracy of the measurements is a function of the sensitivity of the instruments used to make the measurements. For example, a 50-foot tape measure is not as sensitive as a laser measuring device.

- *Variable* data are generated by measuring a specific characteristic against a known standard such as a tape measure or a clock. The results are expressed as fractions or decimals (e.g., a 2 × 4 stud is measured as being 10′6½″ long; a particular model house averages 105.8 days to build). Another example is, "This year the cost to build Model A is $105,678.10, but Model B costs $120,123.45."
- Also known as *continuous* data, variable data are more complex and time-consuming to collect; however, they provide considerably more information about products, services, and process characteristics than attribute data.

5.1.2.3 Other Types of Data

Another way to categorize data is with respect to the number of types of data being evaluated at one time. Sometimes data are referred to with respect to the quantity of related variables:

- *Univariate* data: These data are used when there is only one variable to consider, such as the cost of building a specific model house or the duration expressed in time units (e.g., days or weeks) to construct a community. When only one type of data is being considered, it is treated as being univariate (i.e., one variable). An example of univariate data would be construction cycle time counted in either calendar or working days.
- *Bivariate* data: These data are used when there are two variables to consider, such as the cost of bond exoneration for a community versus time in months until a bond has been exonerated. Another example is customer satisfaction scores (on a linear scale from 1 to 5 or 1 to 7) versus the number of referrals made by home buyers. When two types of data are being considered, they are treated as being bivariate data (i.e., two variables). An example of bivariate data in bond exoneration would

be the number of months to complete bond exoneration as the independent variable and the associated cost per lot of bond exoneration as the dependent variable.

- *Multivariate* data: This is when there are three or more variables, such as the dimensions (length, width, and height) of a room. When more than two types of data are being considered, they are treated as being multivariate data. For example, the strength of concrete (a dependent variable) is a function of multiple factors (all independent variables), such as the amount of water used, the amount and type of aggregate used, and the amount of sand used.

5.1.3 What Is a Variable?

Depending on your source of information, a variable can be defined in many ways, such as the following:

- Something that is likely to vary (i.e., subject to variation)
- Liable to or capable of change (e.g., variable expenses)
- A quantity that can assume any of a set of values
- A symbol such as x or y that is used in mathematical or logical expressions to represent a variable quantity

5.1.4 Types of Variables

There are a variety of approaches to explaining data. This section provides a discussion of the various ways data have been categorized.

5.1.4.1 Variable and Constant Data

Variable data are values that are frequently changing (e.g., basic prices, construction cycle time, lot sizes, and option costs). Constant data are values that never or rarely change (e.g., quantities of doors and windows, interior square footage of a specific home design).

5.1.4.2 Dependent Variables

The value of a dependent variable (e.g., the selling price of a home) depends on the value or size of an independent variable. The price depends on the value or size of a variety of independent variables, such as interior square footage, number and types of options, lot size, lot location, and existing demand for the product.

5.1.4.3 Independent Variables

The size or value of an independent variable is unrelated to the size, value, or presence of any other variable. Examples of independent variables include sales commissions, mortgage interest rates, and closing costs.

5.1.5 Variable Relationships

Relationships between variables are usually expressed in the form of mathematical equations. For example, the cost of building a new home (a dependent variable) can be determined by multiplying the average cost per square foot (a dependent variable) times the size of the home in square feet (an independent constant).

$$(\text{Cost to Build}) = \begin{pmatrix}\text{Average Cost}\\ \text{per Square Foot}\end{pmatrix}\begin{pmatrix}\text{Size of Home}\\ \text{in Square Feet}\end{pmatrix}$$

5.2 DATA ACCUMULATION

Builders that are committed to continuous improvement recognize the importance of data—not just any data but, rather, the "right data." Using the right data to reduce cycle time duration and variation, to improve quality and customer satisfaction, and to minimize costs and accidents is a major step in achieving company-wide continuous improvement. The term *right data* indicates having sufficient current, valid, and accurate data to determine the status of a process and its output.

5.2.1 Predata Collection

Before a builder begins to collect the data needed to assess a process, a product, or a service, a few items should be addressed to ensure that the time and money committed to data collection will not be wasted and possibly require being repeated.

5.2.1.1 Data Stratification

Stratification is the term statisticians (like us) use to describe the categorizing or dividing up of a collection or a group of differing items into subgroups that are not so different. For example, suppose a builder is offering its customers a variety of options composed of various appliances, lighting choices, and landscaping. Now imagine that during the period following closing and move-in, the builder experiences a collection of home buyer complaints. At this point it would make sense to stratify all the options into their component subgroups prior to conducting an analysis to determine which of the option categories has resulted in the most complaints. When this has been accomplished, it would make additional sense to further stratify the identified subgroup; for example, if you assume appliances have the most complaints, then the builder would divide this category into, for example, refrigerators, ovens, cooktops, and dishwashers.

5.2.1.2 Population Consistency/Uniformity

Whenever data are needed to identify and prioritize problems, builders justifiably avoid gathering data from an entire population of homes or home buyers to save time and money. That makes good sense. However, what doesn't make good sense is when a builder uses a sample, that is, a subset of the population, that isn't consistent with the demographics of the population. Examples of demographics for home buyers include annual income and family size and for homes include elevation and plan type. To ensure that the composition of a sample is consistent with the makeup of the population from which the sample is drawn, the demographic percentages in the sample should approximate the corresponding demographics percentages in the population.

5.2.1.3 Sample Selection

Two factors should be considered before a sample is selected:

- First, the units in the sample should be randomly selected from the target population in advance of the data collection process. Randomization can be achieved by using either a published random number table or one generated by specially designed software programs. This topic is addressed in detail in Chapter 10, "Randomization." The application of randomization ensures that the results of a sampling of a population will fairly represent the results that would have been generated had a 100% census been conducted. Randomized sampling has major advantages: It requires less time, costs less, provides fewer opportunities for errors in data collection, and involves fewer personnel, both as data collectors and possible respondents.
- Second, the number of units in the sample should be calculated in advance of the data collection process. The topic of sample size determination is addressed in detail in Chapter 11, "Sample Size Determination."

5.2.2 Data Collection

Data collection is the physical act of counting (for attribute/discrete data) or measuring (for variable/continuous data) specific predetermined characteristics and then accurately recording the results of the inspection.

5.2.2.1 Data Collection Forms (Checklists and Check Sheets)

To ensure that accurate data are easy to record, it is worth locating or creating the right data collection form. Some forms are designed to collect attribute/discrete data, while others are designed for variable/continuous data. Thus, it is important to decide ahead of time what type of data will be collected.

5.2.2.2 Data Collection Procedure

Like so many things, collecting data can be done well or poorly. This is up to the builder. After a sample size has been calculated, randomization has been ensured, and the right data collection form has been selected, then—and only then—is a builder ready to collect data. The person or persons who will collect the data should either already possess the necessary technical knowledge or have received appropriate training to dependably perform this task. The builder should provide ample time and appropriate tools for use by the data collector. These will vary from one type of inspection assignment to another.

5.2.3 Data Sufficiency

People who are not accustomed to working with data accumulation frequently ask the natural question, "How much data do we need?" The answer (a dependent variable) to this question is based on the answers (independent variables) to three other directly related questions. This topic is fully addressed in detail in Chapter 11, "Sample Size Determination."

5.3 DATA ANALYSIS

Once raw data have been accumulated, the next step is to reduce the many bits of collected data into usable statistics. Because there are so many ways to characterize data, this chapter examines some of the most commonly used statistics employed by residential builders.

5.3.1 Measures of Central Tendency

The three most widely used measures of central tendency are (1) the mean (also known as the arithmetic mean or average), (2) the median (the center point), and (3) the mode (the value that occurs most frequently). Each statistic has its own symbol or notation, as follows:

Name	Notation
Mean	$\bar{x}$ (x-bar for a single sample of two or more units)
	$\bar{\bar{x}}$ (x-double bar for a sampling distribution)
	μ (the Greek letter "mu" for a population)
Median	Md
Mode	Mo

Let's use a common data pool to help better understand how to determine each measure of central tendency. Our data pool has five values ($n = 5$): 7, 3, 5, 1, 9. For the sake of discussion, let these numbers be the quantity of defects found by an inspector in each of five homes.

5.3.1.1 Mean

Sum all the values (data points) in the data pool and divide the sum by the number of values, which in this case is the number of homes inspected (the sample size n).

$$\bar{x} = \frac{7 + 3 + 5 + 1 + 9}{5} = \frac{25}{5} = 5$$

5.3.1.2 Median

The median is most often used to represent an estimate of a typical home price. For example, a year ago the median price of a new home in Orange County, California, was $800,000, but now it's down to $600,000, a decrease of 25%. Economists and statisticians prefer to use the median to portray the typical price of a home because the average value can easily be inflated by even one home whose price is considerably greater than the other homes included in a calculation.

To determine the median of a set of values in a data pool, arrange the values into an array of numbers from the largest value to the smallest, or vice versa. The center point is the median. It has as many numbers above it as it does below. This is true with an *odd* number of values in the data pool.

If there is an *even* number of values in the data pool, identify the two values in the middle of the array, and calculate their average value. This average of the two centermost values is the median:

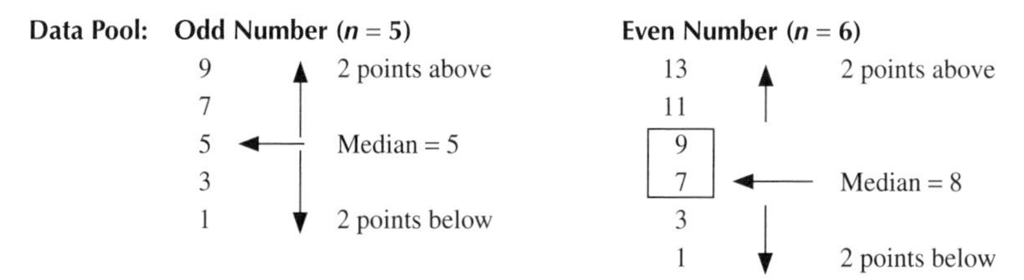

5.3.1.3 Mode

Arrange the values in the data pool into an array of numbers from the largest value to the smallest, or vice versa. The value that occurs most frequently is the mode (unimodal or one mode). If two values occur most frequently, the data pool is bimodal.

An example of the mode used in residential construction would be the most popular option selected from a group of options:

Data Pool:	A ($n = 6$)		B ($n = 6$)	
	13		13	
	11		11	
	11	Mode = 11	11	Mode #1 = 11
	9		9	
	7		9	Mode #2 = 9
	3		7	
	Unimodal		Bimodal	

5.3.2 Measures of Dispersion

The most widely used measures of dispersion (also known as the variability or the variation) are the range, the standard deviation, the quartile, the decile, and the percentile. Data dispersion results from inconsistent, unpredictable performance. A predictable process produces results that are consistent from one day or week or month to the next. An unpredictable process produces results that are widely dispersed from one time period to another with no prior knowledge of when changes to the process are likely to occur. Predictability is important not only to builders, but also to their subcontractors (which need to know when their products and services are actually going to be needed) and their customers (who want to know when their new home will be ready so they can comfortably plan their move out of their old home and into the new one).

5.3.2.1 Range

Identify the largest and the smallest values in a data pool. The range is the absolute difference (without regard to the mathematical sign) between these two values. It is standard practice to subtract the smallest value from the largest value. As a result, the range is always either zero (if the two numbers are the same) or positive.

Data Pool: **($n = 5$)**

7
3
5
1 ← Smallest Value
9 ← Largest Value

$$\text{Range} = x_{\text{Largest}} - x_{\text{Smallest}} = 9 - 1 = 8$$

5.3.2.2 Standard Deviation

The standard deviation (SD) is calculated using this formula:

$$\text{SD} = \sqrt{\frac{\Sigma(x-\bar{x})^2}{n-1}}$$

Data Pool ($n = 5$)

x	$(x-\bar{x})$	$(x-\bar{x})^2$
7	$7-5=2$	4
3	$3-5=-2$	4
5	$5-5=0$	0
1	$1-5=-4$	16
9	$9-5=4$	16
		$\Sigma(x-\bar{x})^2 = 40$

$$\text{SD} = \sqrt{\frac{\Sigma(x-\bar{x})^2}{n-1}} = \sqrt{\frac{40}{5-1}} = \sqrt{\frac{40}{4}} = \sqrt{10} = 3.16$$

$$\bar{x} = \frac{7+3+5+1+9}{n} = \frac{25}{5} = 5$$

5.3.2.3 Quartile

A quartile is 25% of whatever is being evaluated. Four quartiles are used to describe a data pool. The first (least values) quartile and the fourth (greatest values) quartile represent the lowest 25% of all the values and the top 25%, respectively. The second and third quartiles are immediately above and below the median, respectively.

Data Pool ($n = 20$)

Quartile			
1st	**2nd**	**3rd**	**4th**
30	35	40	45
31	36	41	45
32	37	42	47
33	38	43	49
34	38	44	50

Md = 39

5.3.2.4 Decile

A decile is 10% of whatever is being evaluated. Ten deciles are used to describe a data pool.

Data Pool ($n = 20$)	Decile									
	1st	2nd	3rd	4th	5th	6th	7th	8th	9th	10th
	30	32	34	36	38	40	42	44	45	49
	31	33	35	37	38	41	43	45	47	50

Md = 39

5.3.2.5 Percentile

A percentile is 1% of whatever is being evaluated. One hundred percentiles are used to describe a data pool. Thirty percentiles are 30% of whatever is being observed; for example, 30% of $n = 20$ is (30% or 0.30) times (20) = (0.30)(20) = 6 units.

5.3.3 Data Ranking

Pairwise ranking is a step-by-step method for rank ordering a small list of items (usually no more than 10) in priority order. This method was created to help a process improvement team (PIT) prioritize its list as well as to help the team make decisions by consensus. Ranking is important to residential builders because multiple ideas, opportunities, and challenges always need to be prioritized from most to least desirable or costly.

One of the easiest and fastest techniques available for ranking is *paired comparisons.* To begin, a team leader or a group facilitator prints (or clearly writes) the identity of each item on its own slip of paper, preferably with adhesive backing.

Then all the paper slips are placed on a vertical surface such as a whiteboard, window, or wall where all the participants can easily view the ranking activity. Now the team leader or group facilitator starts the paired comparison process by arbitrarily selecting any two slips of paper and placing them next to each other on the same vertical surface separated from but near the remaining slips.

The leader or facilitator then asks the team or group, "Which one of these two items is more (or less) important (or costly)?" Depending on the consensus, the slips are then placed one above the other with the better choice above the lesser choice.

When this ranking is completed, a third slip is selected and compared with the better choice of the previous ranking. Depending on the resulting consensus, the third slip may be placed above or below the better choice of the previous paired comparison. If it is above, this phase is complete. If it is below, then it is time to compare the third slip with the lesser choice of the first paired comparison. Again, depending on the resulting consensus, the third slip may be placed above or below the lesser slip of the initial paired comparison. This completes the second phase.

Next, the fourth slip is arbitrarily selected, and the placement process is repeated. When all of the slips in the original collection have been placed in the ranked sequence of slips, the process is complete. The top slip is the most important, least

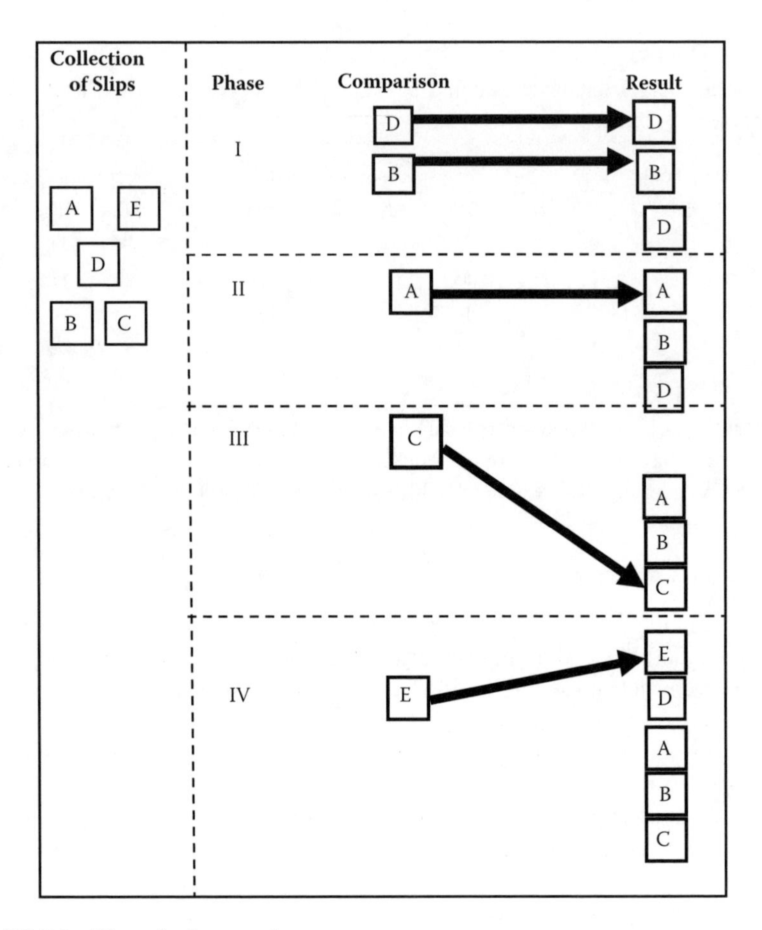

FIGURE 5.1 The paired comparisons process.

costly, or best choice. The bottom slip is the least important, most costly, or worst choice. Figure 5.1 illustrates the paired comparisons process.

5.4 DATA PATTERNS

Data occur in a seemingly endless variety of patterns. The more that is known and understood about any given data pattern, the greater the likelihood that a data analyst will be aware of the consequences of relevant process outputs. The next six sections discuss a variety of common data patterns which are necessary for you to be aware of.

5.4.1 Unimodal Curve

When there is only one obvious peak in a frequency distribution curve (e.g., a bell-shaped curve), the population from which the data were generated is considered to be unimodal. An example of a unimodal population might be the average age of entry-level home buyers.

5.4.2 Bimodal Curve

When there are two apparent peaks in a frequency distribution curve (e.g., a curve with a two-hump, camelback profile), the population from which the data were generated is considered to be bimodal. An example of a bimodal population might be the average income of move-up home buyers.

5.4.3 Multimodal Curve

When there are more than two apparent peaks in a frequency distribution curve (i.e., three or more humps), the population from which the data were generated is considered to be multimodal. An example of a multimodal population might be the average cycle time in days of a range of homes being built that includes entry-level, move-up, and custom homes.

5.4.4 Symmetric Curve

When a mirror held up to the center of a frequency distribution curve reveals a reflection that is the same as the curve hidden by the mirror, the curve is symmetrical; that is, the right and left sides of the curve are the same. A bell-shaped, normal distribution curve is symmetrical. Unless there is something wrong with a saw or its blade, the lengths of a collection of studs that have been cut by a saw will form a symmetrical curve.

5.4.5 Asymmetric Curve

Any time a frequency distribution curve is not symmetrical, by definition it must be asymmetrical. The profile of an asymmetrical curve looks like a bell-shaped curve that has been pushed to one side or the other so that more of the data is displayed on one side of the peak than on the other. When this happens the data are referred to as *skewed*. This is what a builder could expect to see if a saw or its blade were warped or dulled over time.

5.4.6 Other Curves

Numerous other frequency distribution curves may occur from time to time. Builders that believe they have this situation and need some help are advised to contact someone in the mathematics or statistics department in their local community college.

5.5 EXAMPLE

Home builder A has implemented a new customer service program designed to increase customer satisfaction. For the test, they chose two projects with similar product and satisfaction scores and implemented the new process at one while

TABLE 5.1
Results for Both Communities

	New Program	Original Program
1	84.1	81.5
2	87.6	67.9
3	71.5	79.4
4	92.2	83.2
5	88.4	85.6
6	87.1	78.4
7	90.3	91.2
8	91.5	86.7
9	86.9	77.4
10	95.2	72.5
11	91.1	
12	88.4	
13	91.2	
14	89.3	
15	90.4	
16	91	
17	88.7	
18	84.9	
19	90.4	
20	91.7	

keeping the original process at the other. Prior to implementation both communities had a customer satisfaction score of 79.3 with similar standard deviations of 4.5. The results for both communities after implementation of the new program are shown in Table 5.1.

The data were analyzed to determine if the new program made an actual improvement in customer satisfaction. The means and standard deviations were calculated for both the experimental community and the control community and are provided in Table 5.2.

TABLE 5.2
Means and Standard Deviations for Both Communities

	New Program	Original Program
Mean	88.6	80.4
Standard deviation	4.8	6.9

TABLE 5.3
Results of a t-Test

t-score	3.806
p-value	0.0007

In reviewing the descriptive statistics, the new program appears to have made an improvement in satisfactions scores, but to be sure it is not just a random fluctuation in the data, a t-test was performed. The results are shown in Table 5.3.

From the t-test, the probability that the difference between the customer satisfaction scores under the new program and the old program are due to chance is less than 1%. It was therefore concluded that the new customer service program had indeed produced an improvement in customer satisfaction and, therefore, was implemented.

6 Root Causes

6.1 INTRODUCTION

Problem solving is more than just finding a solution on which to get by. Unless the root cause of the problem is addressed, you are making only a temporary or superficial fix.

To illustrate this point, think of yourself as a doctor. A patient comes into your office and describes some of the symptoms that he or she is experiencing. Sometimes there is only one noticeable symptom; other times there are many noticeable symptoms. One option to address the problems would be to treat each symptom individually. Treat the headache with a pain reliever; treat the congestion with a decongestant. The question posed for consideration in this chapter is, "Did we really solve the problem?" Maybe, but it is unlikely. It's unlikely that the patient is aware of all of the symptoms that are manifested by the underlying illness. The patient would not necessarily know if his or her blood pressure was elevated or if his or her immune system was elevated to try to fight off a bacterial infection. In this case, merely treating the symptoms only masks the true problem until the symptoms intensify or additional symptoms manifest themselves. Without going through a proper diagnosis or root cause identification process, we end up only temporarily masking problems instead of truly solving them.

The key is the diagnosis. The wrong diagnosis leads to the wrong treatment. The wrong treatment does not solve the problem. In residential construction, this situation can become apparent in many aspects of the business. For example, when dealing with warranty work, are the customer service representatives just fixing the items home owners have complained about, or are they identifying what aspects of the construction process really caused the problem in the first place so that improved processes can be put in place to prevent or reduce their frequency of occurrence in the future?

6.2 DETERMINATION

To avoid the mistake of treating symptom after symptom individually, we must go through a carefully crafted process of identifying the real source of the symptoms, otherwise referred to as the *root cause*. This chapter provides an overview of the tools most frequently used when facing a problem.

The fishbone diagram, also known as a cause and effect or Ishikawa diagram, is useful in guiding problem solvers in identifying cause and effect relationships with regard to the symptoms being experienced. It is referred to as a fishbone diagram because of its shape, which appears like the skeleton of a fish. The tool is adaptable for use in a broad variety of business settings. Everything from construction defects to administrative problems to customer service problems and many others can be handled by various forms of the fishbone diagram.

The diagram functions by breaking down the symptom or effect into distinct factors to be evaluated as possible sources to the problem. In the case of construction defects, the six factors to be evaluated are as follows:

1. Machines (equipment): Included in this part of the skeleton are possible causes due to machines, tools, or equipment used.
2. Method: Included in this part of the skeleton are possible causes due to the method of construction used.
3. Materials: These are possible causes due to the building materials used in construction.
4. People: Included here are possible causes due to the broad variety of trade persons involved in construction.
5. Measurement: Consideration must be given to potential dimensional errors of duration (time, e.g., years, months, weeks, days, hours, minutes), size (length, width, height), or weight—which could have resulted from problems with accuracy (biased versus unbiased) and/or precision (precise versus imprecise).
6. Environment: Included in this part of the skeleton are possible causes associated with environmental conditions.

If the problem is more administrative or financial in nature, like struggling sales at a community, instead of using the previously mentioned factors we recommend using the eight Ps: price, promotion, people, processes, place, policies, procedures, and product.

This brainstorming technique is qualitative in nature but facilitates the generation of reasonable hypotheses that can be systematically analyzed to find the root cause.

Another useful group of tools that are similar in nature are process flowcharts, process maps, and defect maps. This section focuses on the application of these tools. A detailed discussion on the creation of process flowcharts and process maps can be found in Chapter 13. The process map is the most basic form of this class of tools. It is a two-dimensional, graphical representation of all of the steps, operations, or tasks included in a process. Figure 6.1 offers an example of a process flowchart.

A process map is an expanded version of the process flowchart that illustrates the transfers of both products and services among different organizations, departments, or persons throughout a process. Figure 6.2 shows how a process map differs from a process flowchart. With this tool not only do you see the graphical representation of the steps involved in the process, but it can also be annotated to identify where

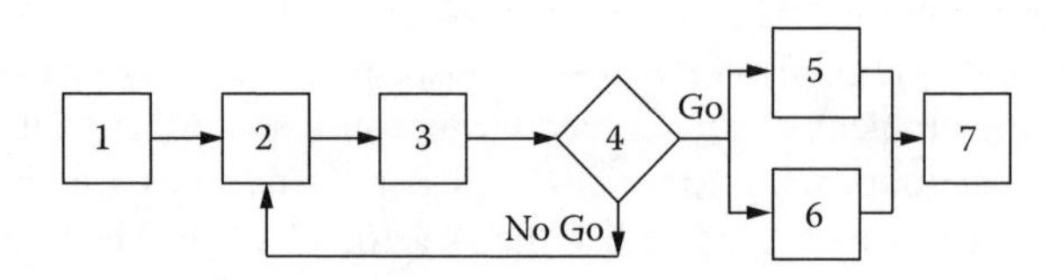

FIGURE 6.1 Process flowchart.

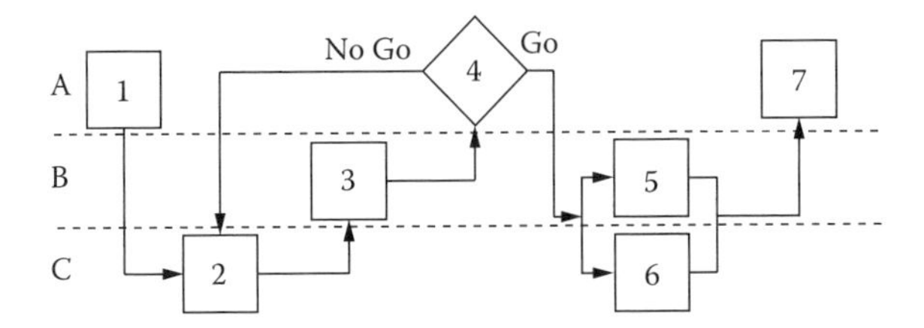

FIGURE 6.2 Process map.

problems are occurring with respect to service flow rates, rework or defects, sources of delay, and accumulation of waste, as demonstrated in Figure 6.3.

In the context of root cause identification, these annotated process maps can be used to identify where and when the problem is occurring within the process. With this knowledge, it is much easier to identify what is truly causing the problem as well as to take timely and appropriate corrective actions.

If the problem you are trying to correct manifests itself as a defect in the product, a defect map can be valuable in determining the root cause. A defect map is a representation of the physical location of defects in the product. For example, if the problem is excessive paint touch-ups, it might be beneficial to map where these touch-ups are needed time after time. If certain areas of a particular home model have high concentrations of paint touch-up work being necessary, it allows for a more focused approach to locating the root cause of the problem. In essence, it can help to identify areas where a problem is consistently recorded as a result of some systemic source because the problem is occurring with much greater frequency in certain specific areas. Figure 6.4 shows a defect map of an entire house.

In some instances, such as analysis of financial data or customer satisfaction data, determining root cause may require data analysis tools such as correlation or regression analysis. These are statistical tools that identify quantifiable

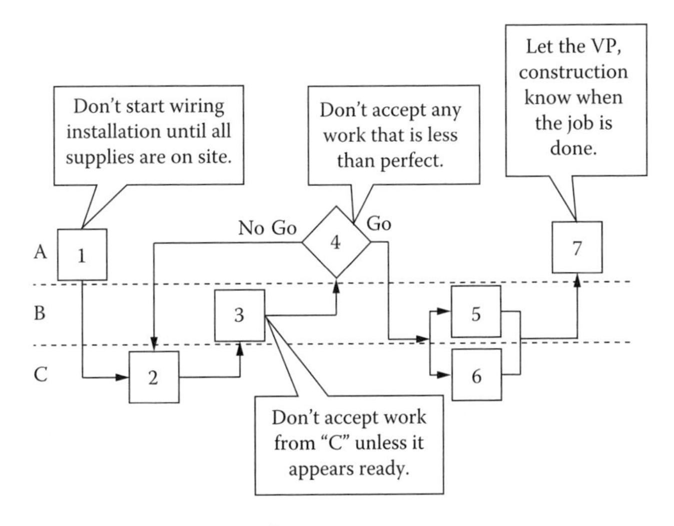

FIGURE 6.3 Process map (annotated).

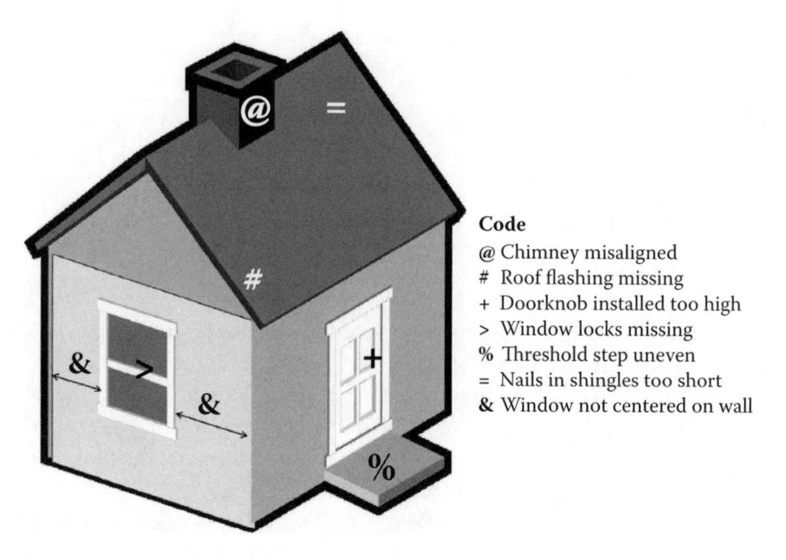

FIGURE 6.4 Defect map of house.

relationships between two or more variables. However, a mathematical relationship does not infer causality. Dealing with improvement of customer satisfaction is an example of such a situation. To improve customer satisfaction, it is important to understand the driving factors of customer satisfaction. While it may seem obvious at first, it is surprisingly difficult to go from simply satisfying your customers to truly delighting your customers. This is because *satisfaction* is an entirely theoretical construct (also referred to as a *latent construct*), created to describe something we cannot directly see, touch, measure, smell, or taste. It can be uniquely different for each individual and each experience. Thus, we use survey questions to attempt to quantify and measure satisfaction, but it is important to realize that these measures are not absolute; they are approximations with varying degrees of precision. Given the nature of customer satisfaction, oftentimes correlation or regression analysis is a necessary first step in identifying what factors are the most influential in determining overall satisfaction. For a detailed discussion of data relationships see Chapter 9.

Another useful tool in determining the root cause of a problem or situation is Pareto analysis, a statistical technique that helps with decision making and uses the 80–20 rule. The 80–20 rule, also known as the Pareto principle after the Italian economist who observed that 80% of the income in Italy in the early 1900s was distributed to 20% of its population, uses the underlying assumption that 80% of the problems can be accounted for by 20% of the causes. While this ratio is not set in stone, its application is still quite helpful. The analysis is accomplished by listing the potential root causes for the problem you are trying to solve and then measuring (counting) the frequency of actual observations. Oftentimes this will lead to one or two primary root causes that account for a majority of the occurrences. Figure 6.5 is an example of a Pareto diagram, and Figure 6.6 shows a multilevel (nested) Pareto diagram.

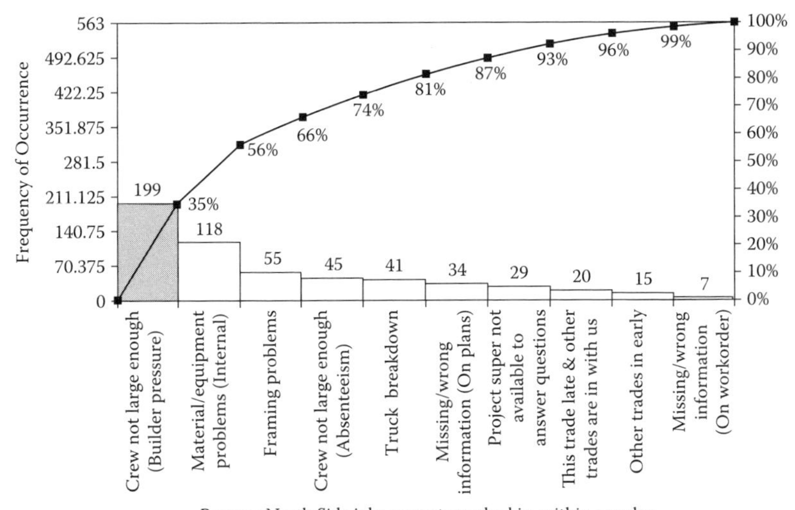

FIGURE 6.5 Pareto diagram.

The list of tools discussed in this section is not exhaustive, but these are some of the most commonly used and most useful tools available depending on the situation. In fact, several of the tools discussed in previous chapters can also be used in identifying root causes. It is important to point out that the tools introduced in this book can be used in multiple stages of process improvement and are adaptable to many different situations.

6.3 SELECTION

With an understanding of the tools and methodologies available, you may be asking, "How do I choose which tool or method to use?" There are few, if any, hard and fast rules about which tools to use when certain things must be taken into consideration. In real-world applications, the most statistically robust method may not always be the best method to employ, especially given the situation. Other considerations such as time, budget, personnel, expertise, and software applications also must be made.

6.3.1 Time

It is no surprise that some data analysis tools and techniques take longer to work through than others. Given the value of time and the implications of deadlines, the time efficiency of the tool or methodology should be taken into consideration in problem solving. This is not to say that shortcuts should be taken; but if a high-level analysis will meet the objective within the time requirements when a more detailed analysis will take too long and cause additional problems, the trade-off should be considered. Illustrations of these time trade-offs can be found within any

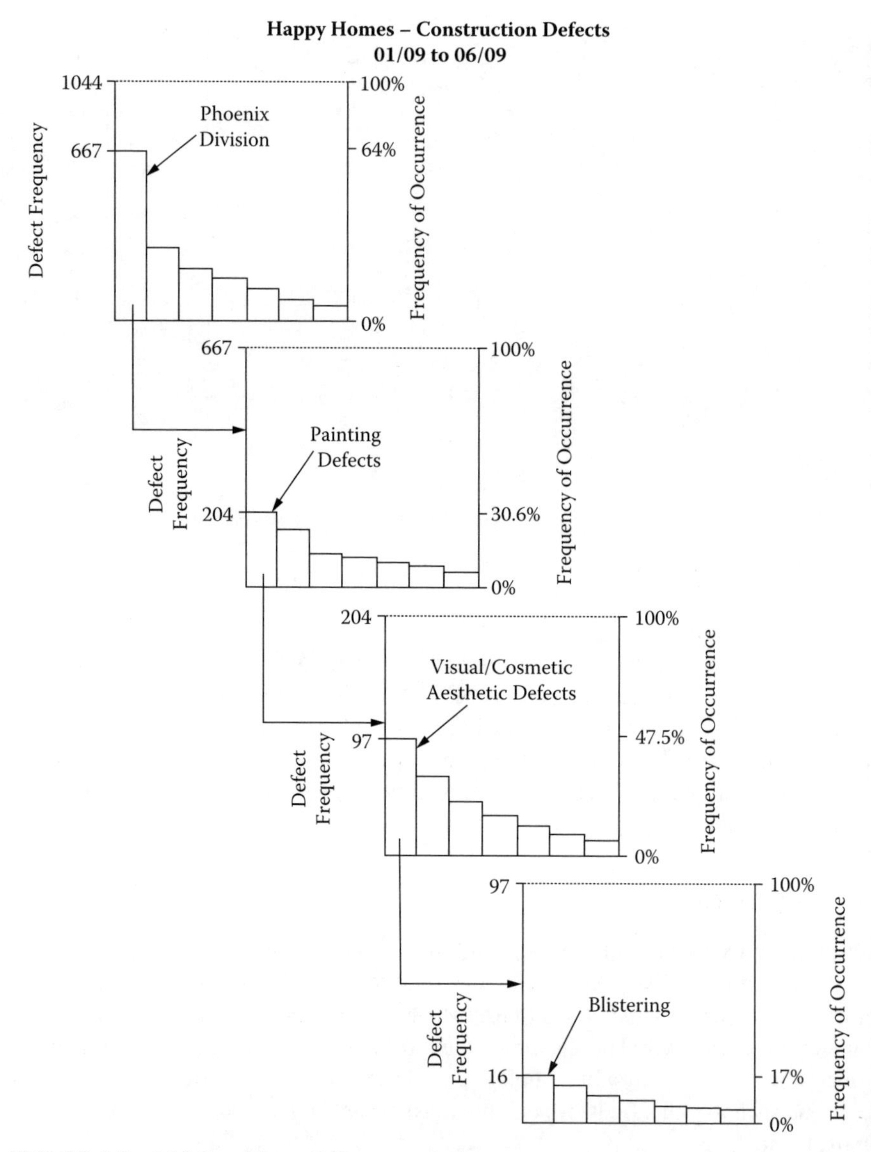

FIGURE 6.6 Multilevel (nested) Pareto analysis.

aspect of the problem-solving process. For example, many companies collect and monitor data on a daily basis; however, the level of detail collected and monitored varies. Some root cause analysis tools require data to be statistically analyzed in an objective manner, while others rely more on experience and firsthand knowledge. If the data needed to perform the statistical analysis are already being captured, the analysis can oftentimes be done as quickly and efficiently as the nonstatistical methods. However, if collection of data not already recorded is required, it could require a substantial amount of time before a useable data set is ready. Under this scenario, if time is a sensitive issue other methods should be used in identifying root cause.

6.3.2 Budget

As with any project, budget should also be taken into consideration when selecting the appropriate tools and methodologies. The goal of all process improvement is to maximize the impact of all resources, especially money. Some tools and methods require a greater investment than others to fully execute. The guidelines for spending will be different for each organization, but care should be taken to avoid spending too much or too little. Overspending on the analysis will diminish the financial impact of the improvement and result in wasted resources, while underspending typically results in failure to adequately improve the process or problem.

In some instances, the situation may call for spending more money than what was budgeted. While it is up to individual managers to assess each situation, it is the general experience and our recommendation that substantially underresourced process improvement projects rarely yield the desired results.

6.3.3 Personnel/Manpower

Another issue is personnel or manpower availability. For example, it is impossible to employ multivoting selection when only one person is voting. It is important to choose the appropriate tools and methods based on the number of people involved in the team. Another example is data collection for regression analysis or Pareto analysis. Are there sufficient people to collect specific data that have not been collected before in a reasonable time frame? If only two or three people are on a team and no other persons can be used on the project, it can be difficult to collect new data for analysis. If the team was addressing a recurring defect in construction, it would be beneficial for the on-site managers to document and collect data as they become available during the scope of the project. If it is up to only two or three people on the team for the entire operation, it would take much more time and effort to collect that type of data. In this instance, it may be better to use data already available or other types of root cause analysis.

6.3.4 Expertise

Two aspects of expertise need to be addressed: representation and capabilities:

- *Representation* refers to the type of expertise possessed by the members of the project team. For example, if you have a team of five people who are all from customer service or all from construction, even though the company may have land development expertise, it is not represented in the project group based on its membership. In most projects it is not necessary to have every department or type of expertise represented, but it is important that those relevant to the problem being addressed be represented.

- *Capabilities*, on the other hand, deal with the depth of knowledge possessed by the members of the group within their expertise. Even if every relevant expertise is represented in the group, there may not be the depth of knowledge necessary to implement certain process improvement techniques or tools. For example, if no one on the team has a working knowledge of how to properly perform and interpret a regression analysis, the team should either find a resource person who does or use an alternative tool. Limiting factors will always be present; the key is to match the method and tools to the team's capabilities.

6.3.5 Software

Some tools and techniques like regression and analysis of variance (ANOVA) require complex computations that are not practical unless accomplished with the use of an appropriate software program. A number of statistical software programs are available that can perform these advanced calculations in seconds, but they do not all have the same functions. It is important to understand what software is needed to perform the right type of analysis for the project. In many cases, programs such as Microsoft Excel will have more than enough capability to implement the tools; however, for more advanced tools, software programs such as Minitab, SPSS, JMP, or SAS are far more effective.

6.4 EXAMPLES

6.4.1 Example 1

Home builder A has seen an alarming increase in frequency and severity of accidents on its construction sites over the past few months. It has become the top priority of senior management to identify and correct the sources of the problem as soon as possible. The first priority of the process improvement team (PIT) is to identify the root cause of the problem.

The PIT starts its task by stratifying the accidents by subcontractor as well as by job or function. This allows the team to identify if it is a particular trade contractor or multiple contractors that are the source of the accidents or a particular task or job that is the major contributor. The results are portrayed in Figures 6.7, 6.8, and 6.9 as Pareto analyses that were performed to select the greatest contributors to the increase in accidents.

The framing companies, Subcontractor F in Figure 6.7, seem to be contributing most to the accidents on the jobsite. The results of the Pareto analysis in Figure 6.8 suggest that the problem is due not to a single company but to a group of companies that all perform the same job function. After identifying framing as the area on which to focus, the accidents were categorized by type. In Figure 6.9 the team identified that a majority of the accidents could be placed into two categories: falls and cuts to feet, bodies, and hands.

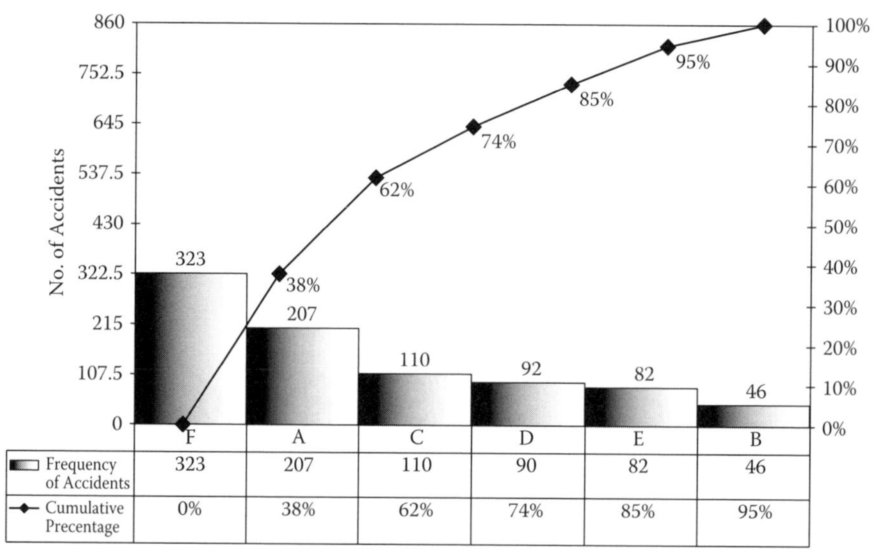

FIGURE 6.7 Pareto analysis of trade contractor accidents: 2009.

6.4.2 Example 2

Home builder B is struggling to complete its warranty repair work when it is scheduled. Frequently, the contractor does not show up with the correct equipment or necessary parts to solve the problem on the scheduled visit, leading to significant delays.

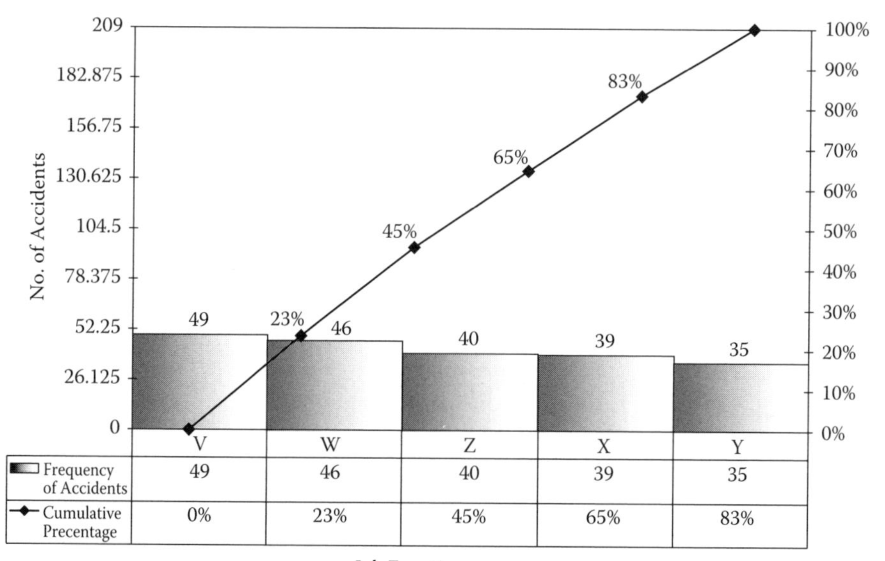

FIGURE 6.8 Pareto analysis of accidents by job function: 2009.

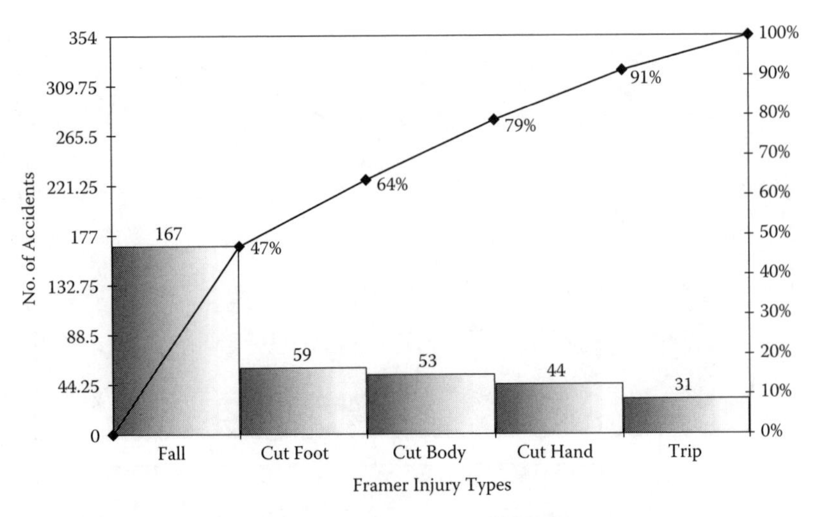

FIGURE 6.9 Pareto analysis of framer injury types: 2009.

This substandard performance has begun to tarnish an otherwise good reputation with potential customers. Something must be done to correct this problem quickly.

Figure 6.10 presents a cause and effect analysis that was performed to select the causes having the greatest impact on the success or failure of the process.

From the cause and effect analysis, the team identified accuracy in communication of work needed and scheduling lead time to be the most significant contributors to the process failure.

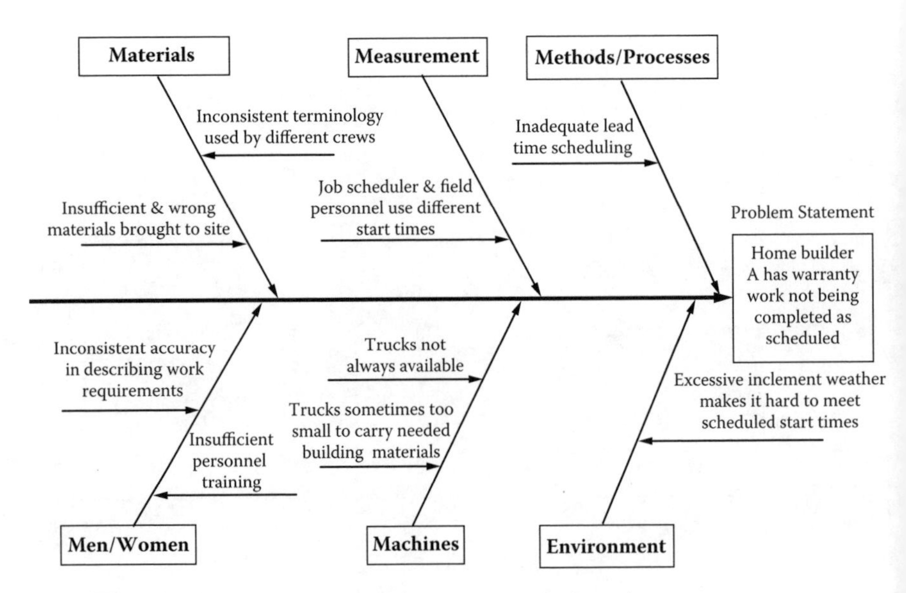

FIGURE 6.10 Cause and effect analysis—warranty repair work.

7 Corrective Actions

7.1 INTRODUCTION

When a problem has been identified and the root cause of the problem has been discovered, it is time to take necessary and appropriate corrective actions. In many of the process improvement methodologies discussed earlier, this phase is referred to as the improvement phase. The objective is to make a change to the existing operation that mitigates or eliminates the occurrence of the identified problem. The word *mitigate* is used because there are some problems that can be minimized but not completely eliminated. For illustrative purposes let's use earthquake damage as an example. A home can be designed to withstand an earthquake of a certain magnitude, but in all reality it is unfeasible if not impossible to design a home to survive the largest earthquake that is theoretically possible. In this situation the design solution implemented is intended to minimize the occurrence of earthquake damage to very rare incidences that cannot be predicted or controlled. In other cases the corrective actions are designed to produce a predetermined goal for improvement based on performance metrics such as reduction of cost or build time.

Regardless of the intended outcome of the corrective action, there are a few important things to keep in mind at this stage. First, the corrective actions should address the root cause of the problem identified. While this seems obvious, it is deceptively easy to fall into this trap. Second, oftentimes there is more than one possible solution to the problem; however, some solutions are better than others. Finally, before corrective actions are implemented globally, they should first be tested and evaluated for potential effectiveness on a smaller scale.

7.2 DETERMINATION

The improvement phase generally starts with a brainstorming session that focuses on solutions to the identified root cause. While a number of brainstorming techniques can be used, the goal at this point is to generate as many potential ideas as possible. Several of the most common forms of brainstorming methods are introduced in this section, and detailed instructions for each form can be found in Chapter 13.

Classical brainstorming is the method most people think of when they hear the word *brainstorming*. It involves a group of people that generates as many ideas as possible for consideration. In this setting there is no such thing as a bad idea. The key is to let your creativity and ingenuity run wild. It is also a group effort, where all members of the group not only try to generate their own new ideas, but also use the other ideas generated as a springboard. One seemingly far-fetched idea can spark another that may ultimately lead to a viable solution.

Brainstorming 635 is a useful form of brainstorming that can be implemented when the dynamics of the team or individual personalities make classical brainstorming less effective. This method of brainstorming is completely nonverbal and is implemented by forming groups of about six people, and they write down three ideas every five minutes. The written ideas are subsequently passed, either clockwise or counterclockwise, to the next person, who then elaborates on what has already been written by adding three more ideas triggered by the ones already on the card. The process is repeated until all the group members receive back their original cards.

Imaginary brainstorming is a technique that has been used to encourage or facilitate the generation of unusual or truly creative solutions to a problem. This is done by breaking the traditional thinking patterns into which people tend to fall. Imaginary brainstorming starts with a round of classical brainstorming around a defined situation. Once completed, one aspect of the original situation is changed, and another round of brainstorming commences. Then the results of the original brainstorm are compared with the imaginary brainstorm, which allows the team to identify more creative solutions to problems.

Antisolution brainstorming is designed to break people out of their normal mode of thinking. This method is quite useful when a team is having a difficult time generating ideas that are not obvious. When using antisolution brainstorming, instead of brainstorming solutions to the problem, the team brainstorms ways to most effectively *cause* the problem. For example, instead of brainstorming how to speed up cycle time, the team would brainstorm the most effective ways to slow cycle time to a crawl. By doing this, possible solutions to the problem are still being identified; it just helps the team look at the problem from a different perspective and to facilitate the generation of new ideas when the group is otherwise stuck in a certain way of thinking.

7.3 SELECTION

While the brainstorming phase fosters the notion that there is no such thing as a bad idea, it must be recognized that some solutions are better than others. The brainstorming session has resulted in a list of potential solutions, and the question becomes: Which solution should be implemented? This is typically answered through a series of debates on the pros and cons of each solution. Each solution will have its strengths and weaknesses; it is up to the team to decide which potential solution is the best given the situation.

In some situations, you may find that one solution is clearly superior to all others. When this occurs, it is usually not necessary to perform a sophisticated analysis of the alternatives; however, even in this situation the solution should be evaluated by asking the question, "Does this solution address and either eliminate or mitigate the root cause of the problem?" If the answer is no, this is not a true solution, no matter how appealing it may be.

This section addresses instances where there are multiple, seemingly viable solutions to the problem, in which three useful tools can aid in selecting the most appropriate one based on the expected impact and the extent of resources and effort required to implement the solution.

7.3.1 Impact–Effort Matrix

The *impact–effort matrix* is essentially a scatterplot that has been divided into four quadrants. Each of these quadrants represents a combination of the amount of effort (high or low) required to implement the solution and the impact (high or low) a given solution is expected to have on eliminating or mitigating the problem. Figure 7.1 contains the template for the impact–effort matrix.

To create the matrix, each potential solution is first rated on a five-point scale for the impact it would have on resolving the root cause of the problem (1 = very little impact; 5 = very high impact). Each solution is then rated using a five-point scale representing the amount of effort and resources it would require to implement the solution (1 = very little effort; 5 = very high amount of effort). After each solution has been rated on both impact and effort, all positions can be plotted on the impact–effort matrix to provide a comprehensive view of each potential solution relative to all the others. Solutions that fall into the low effort–high impact quadrant are solutions that would be easy to implement and have a substantial impact on resolving the root cause. Solutions that fall into the high effort–high impact quadrant are solutions that will have a significant impact on the root cause of the problem but will require a greater investment of time and resources to implement. Low-impact solutions have minimal impact on the root cause of the problem and will provide little value if implemented.

7.3.2 Quantified Force Field Analysis

Another useful tool in evaluating potential solutions is called *quantified force field analysis*. It is more structured and detailed than the impact–effort matrix and provides a visual representation and quantification of potential solutions, but in greater detail. In the quantified force field analysis, each of the factors or forces that potentially promote or hinder the success of a plan, an idea, or an objective is evaluated. When the analysis is completed, each solution will have a corresponding score that can be used to rank order the solutions.

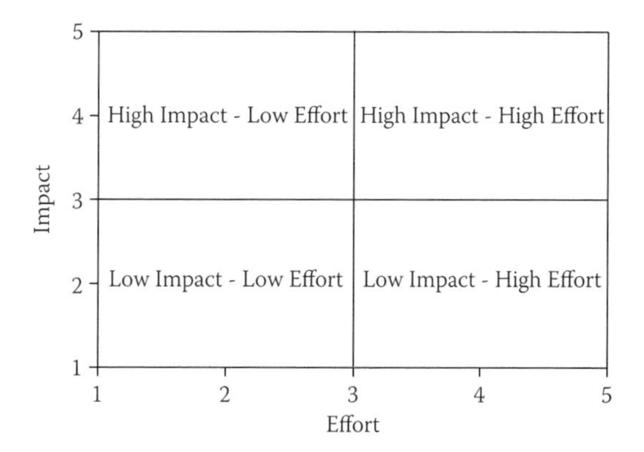

FIGURE 7.1 The impact–effort matrix template.

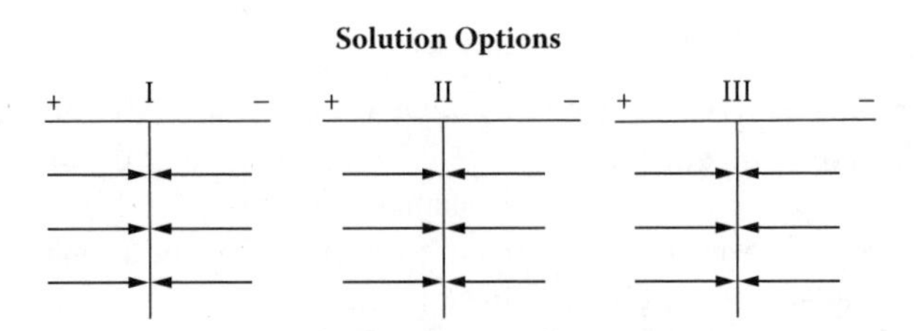

FIGURE 7.2 Template for quantified force field analysis.

The analysis begins with a written statement of the idea, objective, or corrective action plan. A vertical line is then drawn to separate positive and negative forces (factors). Positive forces are generally placed on the left and negative forces on the right. The forces are represented by a horizontal line with an arrow pointing toward the center line. When two forces, one positive and one negative, are in direct opposition, the two arrows should meet at the vertical line and touch. Other forces that are either positive or negative, but are not in direct opposition to another force, should be staggered. Figure 7.2 provides a template for a quantified force field analysis.

After identifying the relevant forces, they are each rated on an importance scale from 1 to 5 (1 = low importance; 5 = high importance). After each force is quantified, all of the values for the positive forces are summed, as are the values of the negative forces. The total score can be obtained by subtracting the sum of the negative forces from the score of the positive forces. Once this is done for every potential solution, the solution with the highest net score would be ranked first, the second highest total score would be ranked second, and so on, as illustrated in Figure 7.3.

7.3.3 Prioritization Matrix

As introduced earlier in Chapter 4 with regard to problem identification, a *prioritization matrix* can also be used to select a solution from several potential solutions. The tool functions in the same manner in both situations. In this case the team develops a list of criteria pertaining to the root cause of the problem that should be addressed by the solution. Each criterion is assigned a weight based on its relative importance. The

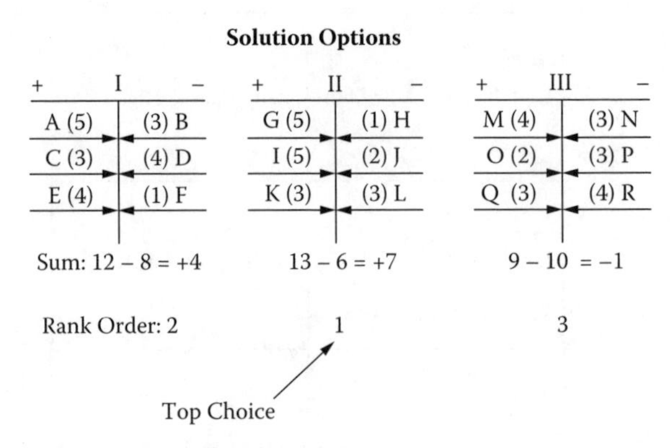

FIGURE 7.3 Example of quantified force field analysis.

Options	Cost (0.26)	Time (0.14)	Resistance to Change (0.01)	Impact on Problem (0.59)	Row Total (1.000)	Rank Order
A	(0.26) (0.31) = 0.081	(0.14) (0.22) = 0.031	(0.01) (0.11) = 0.001	(0.59) (0.29) = 0.171	0.284	1
B	(0.26) (0.12) = 0.031	(0.14) (0.23) = 0.032	(0.01) (0.37) = 0.004	(0.59) (0.19) = 0.112	0.179	3
C	(0.26) (0.12) = 0.031	(0.14) (0.22) = 0.031	(0.01) (0.02) = 0.000	(0.59) (0.27) = 0.159	0.221	2
D	(0.26) (0.33) = 0.086	(0.14) (0.19) = 0.027	(0.01) (0.29) = 0.003	(0.59) (0.04) = 0.024	0.140	5
E	(0.26) (0.12) = 0.030	(0.14) (0.14) = 0.020	(0.01) (0.21) = 0.002	(0.59) (0.21) = 0.124	0.176	4
				Grand Total	**1.000**	

FIGURE 7.4 Example of a prioritization matrix.

team then scores each potential solution against the criterion resulting in a weighted total score. This total score should be used to rank order the potential solutions based on their estimated overall impacts on the designated success factors. Figure 7.4 offers an example of a prioritization matrix.

7.4 IMPLEMENTATION

Once a potential solution has been selected, there is frequently a strong temptation to implement the solution full scale immediately. After all, a team has spent a substantial amount of time and effort in researching, identifying, and developing potential solutions, the result of which will most likely be suitable, right? Surprise—the answer is both yes and no. Yes, the chances are pretty good that the solution derived will have at least a positive effect in the desired direction, but people do make mistakes. There are times when the team will miss a critical aspect of a problem or fail to correctly predict the implications of the solution. These situations will result in the implementation of a solution with less than desirable results. To guard against this, it is important to implement the changes on a small scale first where the results can be observed, measured, and analyzed prior to a full-scale implementation. This is most commonly referred to as a pilot test.

The major difference between a pilot test and full implementation is scale and complexity. A pilot test can be accomplished in either an experimental or applied setting, depending on the problem being faced. For example, if a pilot test is for a potential solution for window leaks, it is probably a better idea to test the solution in a nonproduction environment first. This allows for more rigorous testing of the solution because weather conditions can be simulated instead of waiting for them to happen with real homes. This leads to faster collection of data and reduced exposure if the proposed solution turns out to be inadequate. On the other hand, some improvements can be tested only truly in actual business practice. Things such as modifications intended to improve cycle time can be tested only through building actual homes. In this case, the pilot test program could be implemented at a single subdivision or community. The resulting data can then be compared with previous cycle time data from the same community or for a similar community.

Once the pilot study data are collected, they must be analyzed to determine the effectiveness of the solution. The performance measures used to analyze the solution should include, but not be limited to, the metric used to originally identify the problem. In other words, the analysis must be designed to determine if the solution truly had the desired impact on the root cause of the problem. This concept seems obvious, but if not followed it could lead to expensive consequences.

7.4.1 Statistical Tests

While there is a wide variety of statistical methods and tools that can be used to analyze pilot test data, and far too many to cover every possible situation within the scope of this book, a set of analyses can be applied to a majority of the situations encountered. In most instances the objective is to determine if the data from the proposed solution are statistically different from the original conditions. Depending on the type of data and the parameters of the pilot test, one of the following statistical tests will likely be used.

7.4.1.1 Chi-Square (Test of Homogeneity)

The *chi-square* test is used to test for differences between groups when the metric is categorical, such as pass/fail, go/no-go, and accept/reject. In essence, this test is used to determine if frequency counts are distributed equally between two or more groups. In the window leak example described earlier in this chapter, the resulting contingency table from the pilot test could look as follows:

	Current Method	Proposed Solution
Pass	40	51
Fail	16	8

The chi-square test will determine the probability that the proportions of each solution are actually from the same distribution. In other words, it will help detect if the difference in percentage pass rates is a true improvement or more likely to be just random variation due to sampling.

To compute the χ^2 (read "chi-square" and pronounced "kai-square") statistic, the following formula should be used:

$$\chi^2 = [(ad) - (bc)]^2 \times (a + b + c + d)/(a + b) \times (c + d) \times (b + d) \times (a + c)$$

The variables correspond to the following contingency table:

	Current Method	Proposed Solution	Totals
Pass	a	b	$a + b$
Fail	c	d	$c + d$
Totals	$a + c$	$b + d$	$a + b + c + d$

Using these data in the chi-square formula yields the following result:

$\chi^2 = [(40*8) - (51*16)]^2 \times (40 + 51 + 16 + 8)/(40 + 51)(16 + 8)(51 + 8)(40 + 16)$
$\chi^2 = [320 - 816]^2 \times (115)/(91)(24)(59)(56)$
$\chi^2 = [496]^2 \times (115)/(91)(24)(59)(56)$
$\chi^2 = 28{,}291{,}840/7{,}215{,}936$
$\chi^2 = 3.9207$

After obtaining the chi-squared statistic, the degrees of freedom (df) must be calculated using the following formula:

$$df = (\text{Number of columns} - 1) \times (\text{Number of rows} - 1)$$

In this case our degrees of freedom $(df) = (2 - 1) \times (2 - 1) = (1)\,(1) = 1$.

We can now use the following chi-square distribution table to determine if the difference between our current method and the proposed solution is statistically significant:

	Probability Level or Alpha Level					
df	$p = .5$	$p = .1$		$p = .05$	$p = .01$	$p = .001$
1	0.455	2.706	⟷	3.841	6.635	10.827
2	1.386	4.605		5.991	9.21	13.815
3	2.366	6.251		7.815	11.345	16.268
4	3.357	7.779		9.488	13.277	18.465
5	4.351	9.236		11.07	15.086	20.517

Using the chi-square statistic of 3.9207 from our calculation with 1 degree of freedom, we find that it falls between the values in the table that correspond to p-values of .05 and .01 (Noted in the table by the two-headed arrow.) The p-values, in turn, correspond to the probability that the difference between the chi-square statistic and the tabularized (control) statistic is actually due to random sampling error. These data indicate that the probability that there is no real difference between the current method and the proposed solution is less than 5%. This would generally be considered statistically significant and interpreted as meaning that the proposed method is an improvement over the current method.

7.4.1.2 t-Test (for Independent Samples)

The t-test is probably the most widely used statistical test available. It is designed to evaluate the difference in averages (means) between two groups. For example, if the objective is to reduce cycle time, the t-test for independent samples could be used to evaluate the difference between the average cycle time of the experimental group and the average cycle time of the comparison or control group. The latter might be the average cycle time of homes built in the same subdivision prior to the change, or it could be the average cycle time of homes built in a comparable subdivision. In theory, the t-test can be used with very small sample sizes. While some statisticians have argued that the t-test can be used on sample sizes as small as 10, it is generally recommended that a sample size of at least 20 be used. If obtaining an appropriate sample size is a significant limitation or even impossible, equivalent nonparametric tests such as the *Wilcoxon t-test* or the *Mann–Whitney U-test* can be used. These tests are more robust (insensitive) for smaller sample sizes because they have no assumptions of underlying distributions and are therefore considered distribution-free statistical tests. These common tests are available in most statistical software packages such as the ones described in Chapter 12, but can also be easily computed by hand for small sample sizes.

The t-test for independent samples is computed using the following formula:

$$t = \frac{\bar{x}_1 - \bar{x}_2}{S_{x_1 x_2} \bullet \sqrt{\frac{2}{n}}}$$

The notation in the numerator represents the average of Sample 1 minus the average of Sample 2. In this chapter's example this would be the difference between the average cycle time of the test group and the comparison (control) group. The notation in the denominator is the square root of 2 divided by the number of observations multiplied by the pooled standard deviation of Samples 1 and 2.

The pooled standard deviation can be estimated using the following formula:

$$S_{x_1 x_2} = \sqrt{\frac{S_{x_1}^2 + S_{x_2}^2}{2}}$$

7.4.1.3 Analysis of Variance (ANOVA)

Analysis of variance is similar to the t-test since it is used to determine if the averages between various groups are statistically different. The distinction lies in the fact that a t-test can be performed only on the difference between two groups, whereas an ANOVA can be used to compare multiple groups. Assuming the appropriate samples size and design are available, there is really no limit to the number of groups that can simultaneously be tested using ANOVA. A detailed discussion of ANOVA can be found in Chapter 15, "Design of Experiments," as well as many other statistics books.

After analyzing the results of the pilot study, it is time to make the decision to either implement the solution full scale or to revise the solution and try again. If the solution tested had a positive impact on the problem but not enough to meet the requirements, the team may look at how to improve the proposed solution to make it more effective. If the tested solution failed completely, the team should then proceed to pilot test the highest-ranking alternate solution.

With all of the different tests available, it can be a bit daunting to decide which test to use for each situation. Figure 7.5 presents a decision tree for statistical analysis tests that can be used as a guide in choosing the appropriate analysis technique. The first question that must be asked is, "What type of data requires analysis?" If the response is categorical (e.g., the quantity of each of three different model homes),

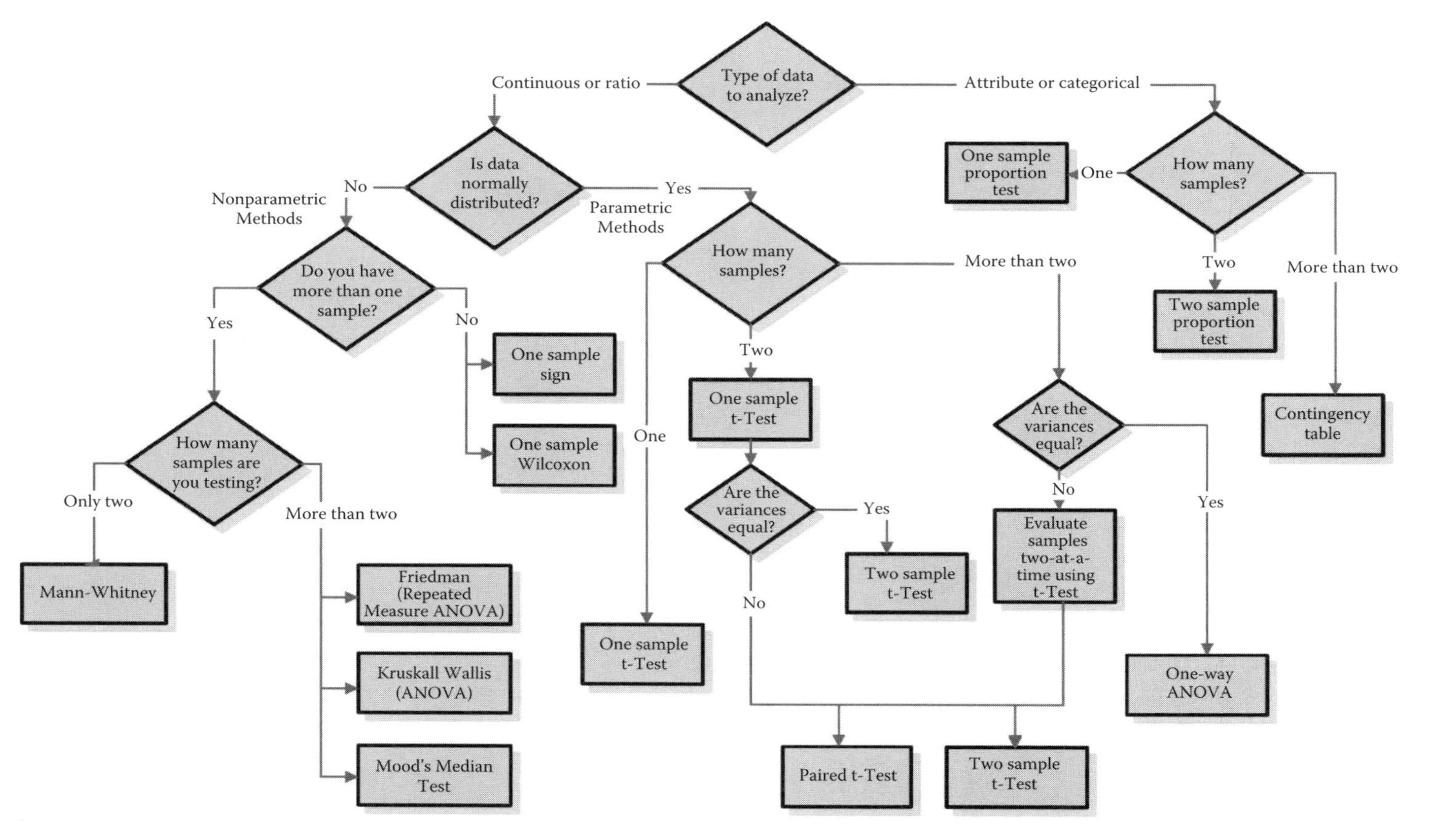

FIGURE 7.5 Decision tree for data analysis.

it will lead to one series of analysis; if the answer is continuous or ratio data, it will lead to another set of options.

7.5 EXAMPLE

Builder A has identified the root cause of its increase in accidents as being attributed to framers falling and eye injuries. The team has brainstormed several possible solutions to both problems ranging from increased training programs to increased supervision and physically altering the equipment to prevent further injuries. A force field analysis was performed to select the optimal corrective action when considering the various positive and negative factors associated with each potential corrective action.

After implementing the solution at a few projects, the pilot test data were analyzed to determine if the solution has produced the desired effect using a t-test on accident severity and frequency of occurrence. The resulting statistical test indicated that an improvement had been made and that the solution is ready for full-scale implementation.

8 Problem Follow-Up

8.1 INTRODUCTION

After pilot testing the solution to a problem and then deciding to implement it globally, the hard work is essentially done. However, this does not mean the job is complete. The term *continuous improvement* means exactly what it states: The job is never complete. After implementing the solution, a follow-up procedure should be implemented to monitor the improvement. In the Six Sigma methodology, this phase is referred to as the *control phase.*

The first step is to develop a process monitoring plan. It does not have to be an overly complex or sophisticated process; the objective is to develop a performance metric that can be tracked on a consistent basis and that will allow for detection of future problems before they become widespread. Oftentimes these metrics are what brought attention to the problem in the first place, but occasionally it's necessary to develop a new metric that is more closely associated with the performance of the newly improved process. For example, if a problem was initially identified by the presence of unusually high construction costs and the root cause was identified as being the necessity for extensive concrete rework, it would be advisable to develop a metric to track the improvement in concrete rework rather than relying on the overall construction cost metric, because the latter includes all other aspects of cost, not just concrete rework. If the overall cost metric were to increase again, there is no way of knowing whether it is due to the improved concrete process failing again or to some other factor. By developing a monitoring metric on the improved process specifically, management will be aware how well the improved process is functioning. If another problem is indicated by the overall metric, it will be easy enough to determine if it is due to some new cause or perhaps a failure in the improved process.

The next step in the problem follow-up phase is process standardization. When the solution is implemented, it should be implemented in a standardized form across the company. Standardization helps ensure proper and complete implementation as well as predicable results. Process standardization should include all the necessary documentation for the process. Documentation serves as a reference guide for implementation across multiple divisions in the case of a medium-size to large company or across multiple communities in the case of a small company as well as assurance that the knowledge is not lost over time through employee turnover. In addition, fully documented processes can be easily reviewed for potential improvements when new technologies become available.

At some point, a transfer of ownership needs to take place from the problem-solving team to the employees and managers who are responsible for the ongoing operations. If the problem was construction related, the ownership would be transferred to the field superintendents and construction managers. If the problem was part of an administrative process in accounting, the ownership needs to be turned back over to the accounting leaders. Without this transfer of ownership, the implemented solution may or may not stay in place. This failure of a solution methodology is frequently simply a lack of communication between the project team and the appropriate managers. If the responsible managers and employees do not know why a process was changed, the solution methodology will be susceptible to being ignored, changed, or implemented incorrectly. The managers who are directly impacted need to know that they are now in charge of maintaining the new process as well as collecting and monitoring the data.

Depending on the goals initially established, by this time the process improvement team should have identified whether additional solutions are needed to achieve the goal. If the goal was to reduce cycle time by 25% and the initial solution achieved an 18% reduction, an additional solution would be required to meet the original goal. With this, the process essentially loops back to the beginning, and the team identifies the next steps and plans for the remaining opportunities and potential solutions.

8.2 DATA COLLECTION AND ANALYSES

Data collection and analysis in the problem follow-up phase typically fall into the discipline of statistical process control (SPC). Again, this book does not attempt to cover the entire field of SPC but instead focuses on the most common and useful tools for process follow-up.

The first step is to identify the key input variables to be monitored. For example, to monitor cycle time reductions, the average cycle time could be used. Alternatively, if you are monitoring an improvement in safety, your key input variable may be accident frequency or severity. In any case, if the wrong metric is selected, no type of statistical test or tool will be able to make up for it. As with any analysis, planning prior to analysis is incredibly important. The metric used to indentify the problem or to analyze the potential solutions often will be the same metric used in the follow-up phase, but not always. The metric should always be as closely related as possible to the root cause of the problem solved.

One of the most commonly used tools in the field of SPC is the control chart. Control charts are designed to help quickly and easily identify real-time process changes or deviations from acceptable tolerances. This seemingly simple decision can be much more difficult to make in practice when the metric being monitored has a high degree of variation. The benefit of the control chart is that it applies objective criteria and decision rules to determine when a process is in or out of statistical control. Having a process in statistical control is synonymous with the process being predictable within a known, calculated range—the range being from the lower statistical control limit to the upper control limit. The result is a minimization of false alarms or overreaction to random fluctuations in the data and quick detection of true process shifts or problems. It is worth pointing out that control charts can detect both positive and negative changes. When a positive change is detected, the process can be analyzed to determine what caused the improvement so it can be standardized and replicated. Alternatively, when a negative shift is detected, the process improvement team can institute appropriate problem-solving methods or follow-up procedures already in place.

While they are all based on the same general concept of detecting statistical deviations from an expected value using upper and lower control limits, there are many different types of control charts designed to fit the type of data being used and the process to be measured. The following sections provide a brief overview of the types of control charts available and under what conditions they should be used. For a more in-depth review of control charts and their mechanics, a number of statistics books and websites provide further detail. Additionally, most statistical software packages will perform all the necessary computations, so the most important thing is to understand how the control charts are constructed and what they are designed to do. The control charts are generally used to track measures of central tendency, such as an average or proportion, while others track measures of variation, such as range or standard deviation. A common mistake is to rely only on the measure of central tendency and to ignore the measure of variation. Since metrics like the average provide only part of the story about what is happening in the data, relying solely on that single metric can result in costly errors due to faulty assumptions. This is illustrated in further detail in the context of process capability.

8.2.1 X-Bar & R and X-Bar & S Charts

These charts are used to monitor the arithmetic mean (also known as the average) of successive samples of equal size from a process over time. Examples include average cycle time, average construction cost, or average severity of safety incidents. Figure 8.1 provides an example of an X-bar and R chart.

The X-bar chart contains three primary components of information. First is the center line, which is derived from either the historical or the expected process

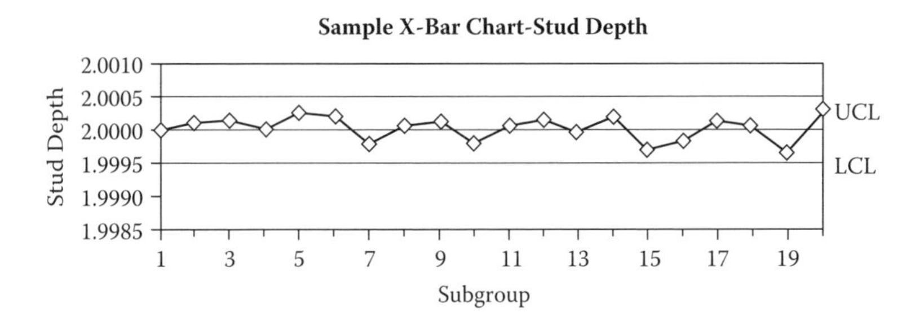

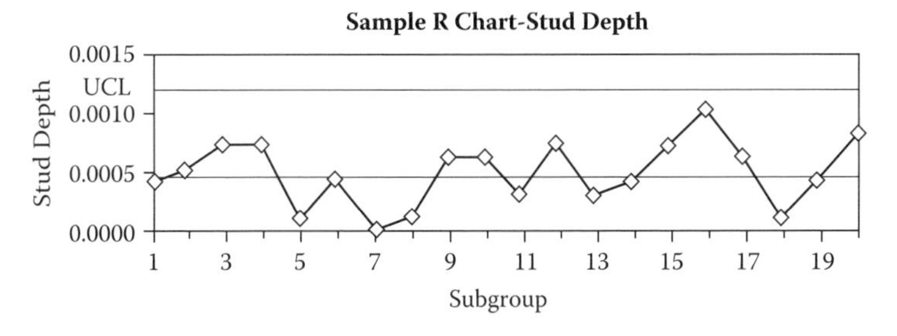

FIGURE 8.1 Example of X-bar and R chart.

average. The second component is the upper and lower control limit boundaries. These boundaries are derived from the historical variation in the mean, also known as the standard error. The upper boundary is typically placed three standard error units above the expected mean, while the lower boundary is placed three standard error units below the mean. The third component is the actual sample means. Every sample is plotted in relation to the expected mean and the control limits. The X-bar chart is typically used in conjunction with either the R chart or the S chart because the X-bar chart tracks variability between separate samples while the R and S charts track variability within each sample. The S chart tracks the standard deviation of the data for the corresponding samples with upper and lower control limits based on the same principles as the X-bar chart. The R chart is a simplified version of the S chart and approximates the variation in the data using the range of scores as opposed to the standard deviation.

8.2.2 Individual and Moving Range (IMR) Chart

The IMR chart (also known as the XmR chart) is similar to the X-bar chart with the exception that instead of sample data points being averages of multiple observations, each successive data point is the result of a single observation. This type of chart could be used to track cycle time at individual subdivisions with slow absorptions or tracking individual floor plans. It is constructed based on the same theory as the X-bar chart, but the control limits are calculated using an equation based on the moving range rather than standard error. Since the IMR chart incorporates the variation into the calculation of its control limits, it can generally be used as a stand-alone chart.

8.2.3 P Chart

The P chart monitors the proportion of nonconforming units in a sample. It is typically used when the data collected are in terms of pass versus fail. Quality inspection data on specific components of the home could be used in a P chart. The data collected are turned into a proportion or ratio.

8.2.4 Np Chart

The Np chart is similar to the P chart but instead of using the percentage of nonconforming units, it uses the actual count data. Given the type of data, this chart does not calculate the control limits using the normal distribution as with the X-bar chart, rather it uses the binomial distribution.

8.2.5 C Chart and U Chart

Sometimes it is necessary to quantify the number of nonconformities or defects per unit rather than to classify an entire unit as pass or fail. In this situation, the C chart and the U chart are the control charts designed for this situation. In residential construction, these charts can be applied to the number of defects per home at specific inspection points. Instead of classifying the entire house as pass versus fail, the total number of items or the average number of nonconformities per unit can be tracked. In short, it allows for multiple nonconformities per unit. The C chart is designed for sample sizes of one or more sequential homes being constructed. The major difference between the C chart and the U chart is the sample size assumption. The C chart requires a constant sample size, whereas the U chart allows for varying sample sizes.

8.2.6 Other Charts

Some instances require more advanced control charts to detect shifts in the process away from the desired or expected average. For example, the control charts in the previous section are very good at reacting to large shifts in the data away from the mean or expected value but are quite slow to react to gradual or small shifts. If detecting small or gradual shifts is a high priority for a project, two types of control charts are more sensitive and better suited for your needs. The exponentially weighted moving average (EWMA) and cumulative sum (CUSUM) charts are designed to detect just that. While it is outside the scope of this book to go into detail on the mechanics of these charts, it is important to know which charts are available and best suited for your needs.

It can be a bit daunting to select the proper chart to use if you are not involved with making this type of decision on a regular basis or haven't spent a number of years studying statistical theory, so Figure 8.2 provides a decision tree to guide you to the correct control chart for your application.

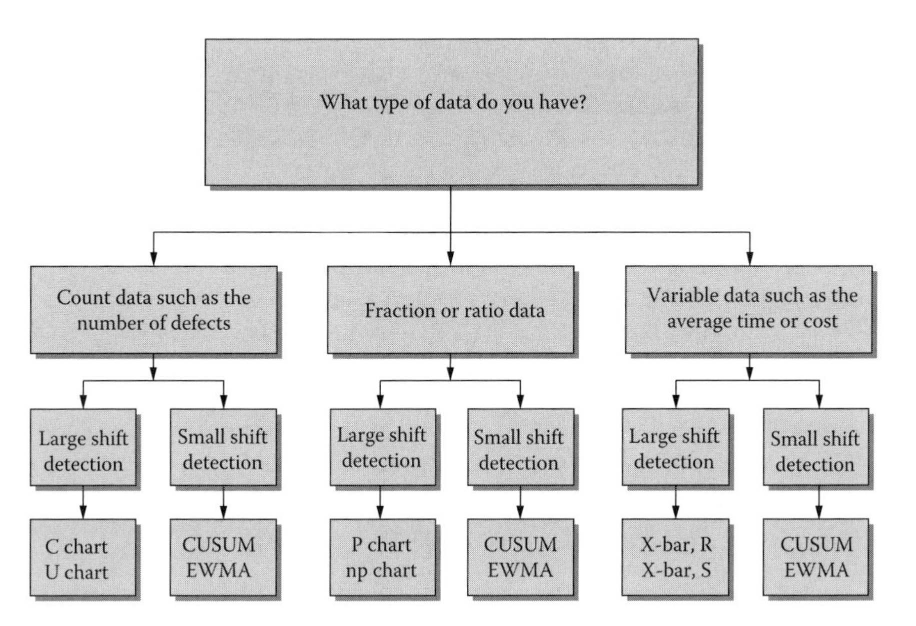

FIGURE 8.2 Control chart selection tree.

8.2.7 Shift Detection

Once a control chart has been selected and developed, how can it be determined if the monitored process is in control? A number of signals point to a shift away from the expected mean; therefore, it is an out-of-control process. The first and most obvious method of detection is through use of the control limits. If a sample point occurs outside either the upper or lower control limits, the chart has detected a significant change from normal operations. Other methods involve specific patterns of consecutive points. For example, if six consecutive points are found to be increasing or decreasing or seven consecutive points occur above or below the center line without crossing over, it is likely that a shift has occurred and the process is no longer in control.

8.2.8 Process Capability

While control charts can help determine if a process is in statistical control, it does not necessarily provide all of the information necessary to determine if a process is performing to customer or business standards or requirements. A process under statistical control means only that it is in a predictable state. To determine how well a process is meeting requirements, process capability metrics were developed (C_p, C_{pk}). Process capability quantifies the percentage of units produced within user-defined specifications. The important difference here is that the limitations are not statistical boundaries but rather are company (internal) or customer (external) specifications. To illustrate this point, think about building codes. These are specifications pertaining to construction that are independent of the construction process and have a significant impact on the product itself. If a home is not built to code, it is considered defective and becomes a significant liability to the company through repair work and possible litigation. Without getting bogged down in the individual codes for each state or municipality, let's use walls as an example. Regardless of whether the walls are framed with wood or made of block, they are required to be plumb (perpendicular) to the foundation; however, there is a tolerance level associated with that metric. It is permitted for walls to be slightly out of plumb, but only to a certain defined deviation. If the walls are randomly sampled and measured and the average deviation is within the acceptable tolerance, it does not mean that there is not a problem. It just means that on average the walls are within tolerance, but that a percentage of the walls will not meet specifications. Figure 8.3 illustrates the difference between a stable, capable process (on the right side) and a process that is unstable and not capable (on the left side), but both have the same average score.

Process capability is estimated using the following formulas. If a process is centered, meaning the performance metric being evaluated is midway between the upper and lower specification limits, process capability can be assessed using the following equation, where:

- C_p represents process capability
- USL and LSL are values for the upper and lower specification limits
- 6σ stands for six times the standard deviation of the measured process

$$C_p = \frac{USL - LSL}{6\sigma}$$

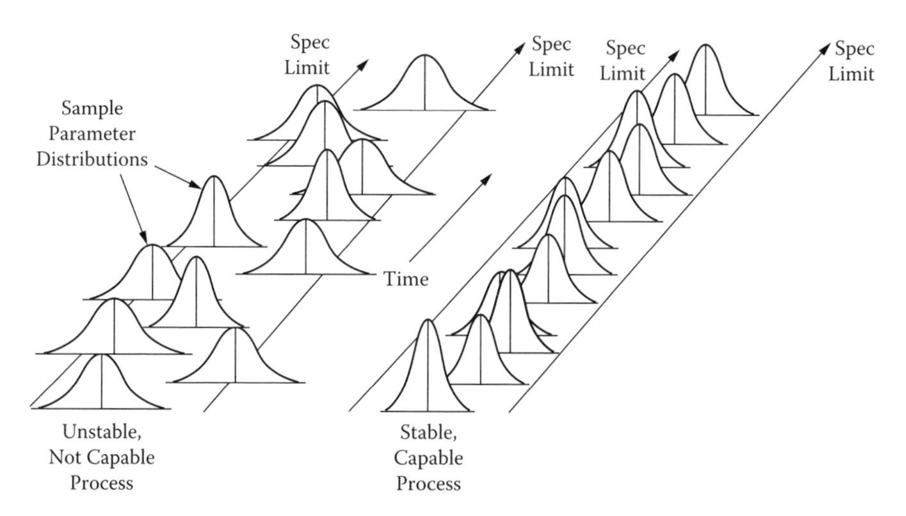

FIGURE 8.3 Capable versus not capable processes.

If a process is off center, the equation is modified to use the smaller value of the process capability of the specification limit closest to the process average. In the following equation, a capability index can be calculated for the upper specification limit as well as the lower specification limit by dividing the distance between the average and the specification limit by three standard deviations. The process capability is the lesser of the two values since the process is generating more defects with respect to the spec limit that is closer to the process average. Figure 8.4 compares a noncentered process on the left side with a centered process on the right side. In the following equation, μ (the Greek letter *mu*) is the symbol for the process average:

$$C_{pk} = \min\left[\frac{USL-\mu}{3\sigma}, \frac{\mu-LSL}{3\sigma}\right]$$

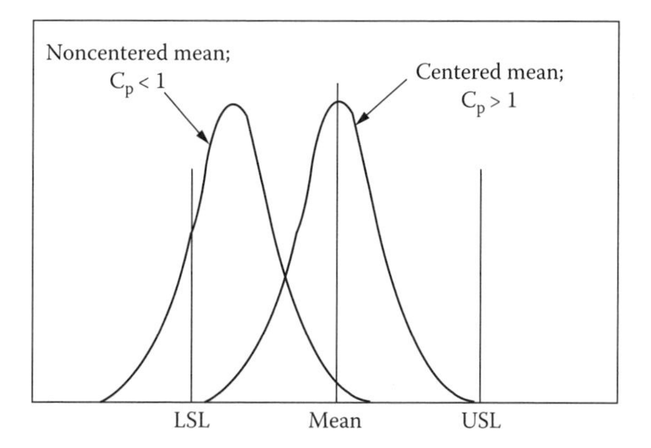

FIGURE 8.4 Centered versus noncentered processes.

TABLE 8.1
C_{pk} versus ppm

C_{pk}	Sigma	ppm
0.43	-	193,600
0.47	-	161,510
0.50	**1.50**	133,610
0.53	-	109,600
0.57	-	89,130
0.60	-	71,960
0.63	-	57,430
0.67	-	45,500
0.70	-	35,730
0.73	-	27,810
0.77	-	21,450
0.80	-	16,400
0.83	-	12,420
0.87	-	9,322
0.90	-	6,934
0.93	-	5,110
0.97	-	3,732
1.00	**3.00**	2,700
1.03	-	1,935
1.07	-	1,374
1.10	-	967
1.13	-	674
1.16	**~3.50**	465
1.20	-	318
1.23	-	216
1.27	-	145
1.30	-	98
1.33	**4.00**	64
1.37	-	41
1.40	-	27
1.43	-	17
1.47	-	11
1.50	**4.50**	7
1.53	-	4
1.57	-	3
1.60	**~5.00**	0.5
2.00	**6.00**	0.00198

The process capability score can then be transformed into a parts per million (ppm) defect opportunities value, which indicates how often defects are occurring and will occur using the current process. Table 8.1 contains mathematical equivalencies between C_{pk} and ppm.

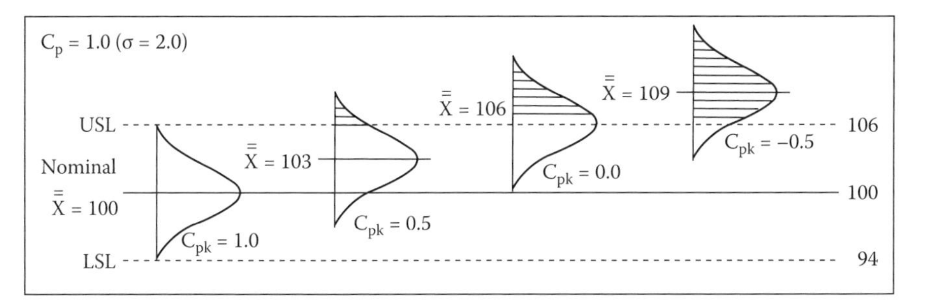

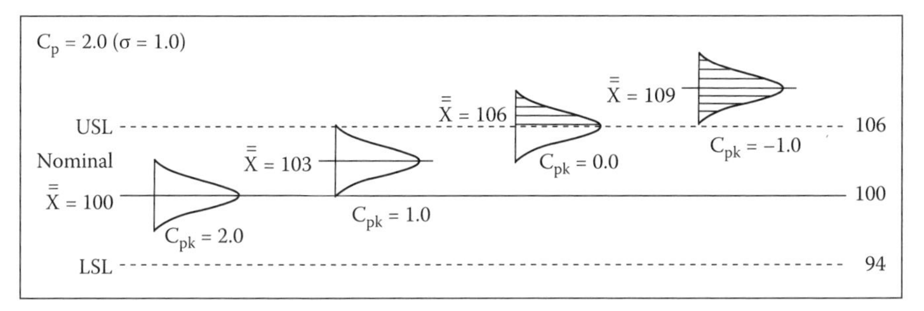

FIGURE 8.5 C_p and C_{pk} relationships.

It is important to remember that a process can be capable only if it is already in statistical control (predictable). If a process is not in control, the capability metrics are useless because the process capability (C_p) would vary from one time period to another. Sometimes it would produce many defective outputs while other times it could produce only a few. When a process is not performing in a predictable manner, the first priority must be bringing it into control before attempting to address capability. It is also entirely possible for a process to be in control but not capable. In this case, expect the process to consistently produce defective outputs.

To further clarify this situation, Figure 8.5 demonstrates a comparison between a process (at the top of the figure) with a C_p of 1.0 (this is equivalent to 3σ or 2,700 ppm) and a standard deviation (σ) of 2.0 while the process average is increasing from left to right (and the corresponding C_{pk} is deteriorating from 1.0 to –0.5) with another process (at the bottom of the figure) with a C_p of 2.0 (this is equivalent to 6σ or 3.4 ppm) and a standard deviation (σ) of 1.0. It should be noted that the bottom process results in far fewer defects (the shaded area under the normal distribution frequency curve outside the statistical control limit). It should be further noted that when the process deteriorates sufficiently, even a C_p of 2 (or greater) won't prevent the occurrence of a massive number of defects.

Whenever comparing process capability indexes, three important rules must be noted:

1. When $C_p \geq 1.0$, it can be concluded that the process of interest is capable of producing product within the required specification limits.

2. When $C_{pk} \geq 1.0$, it can be concluded that the product produced by the process of interest fits within the required specification limits.
3. Even when $C_p \geq 1.0$, C_{pk} must also be ≥ 1.0 to be certain that a capable process is producing acceptable product.

8.3 PROCESS MODIFICATION

The follow-up process also starts the next iteration of process improvement. With a standardized process that is performing at a known level of capability, the next step is to loop back to the beginning to try to improve the process even further. It does not have to be immediate by any means, but don't get complacent with the current performance. Sometimes it will be a new idea about improving the process that had not yet been thought of, or it could be that an old solution previously determined to be impossible to implement due to some sort of technological or economic constraint has become a viable opportunity due to changes in the industry or business environment. As technology changes, it also provides new opportunities to improve current processes that are performing at reasonable levels. As stated earlier in this book, the key is the systematic approach to improvement. Building on a body of knowledge to improve the process as new opportunities present themselves rather than random or haphazard attempts are the keys to real sustainable improvement.

8.4 EXAMPLE

After pilot testing the safety enhancement program, it was rolled out company-wide. With the rollout of the program also comes the work of setting up the monitoring system. The process improvement team (PIT) has decided to monitor both the frequency of accidents and the severity of accidents to ensure the process is performing

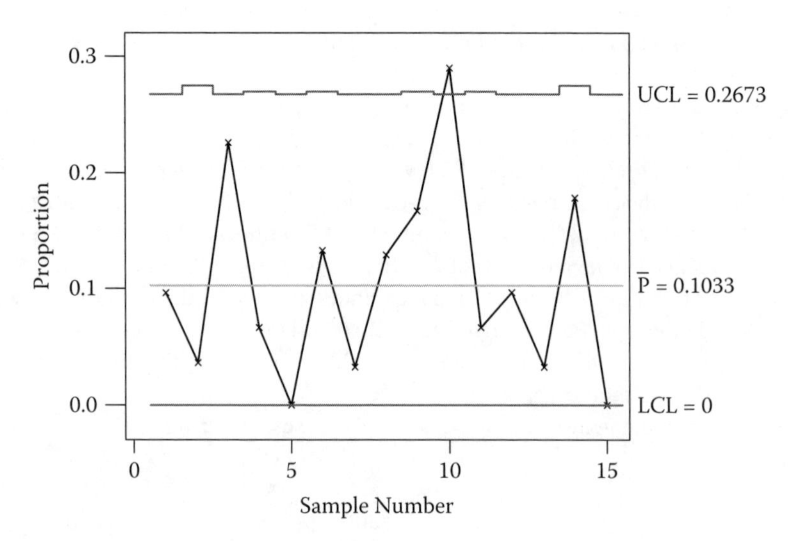

FIGURE 8.6 P chart for accident frequency.

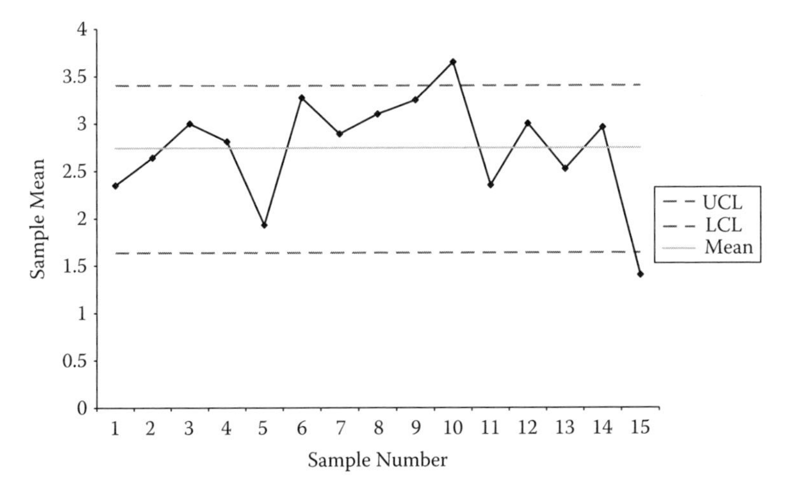

FIGURE 8.7 X-bar chart for accident severity.

to standard. The frequency metric is defined as the number of accidents recorded during a specified time. Because these are count data, and it is difficult to estimate the total number of accident opportunities to turn the count into a ratio, the team decided that using a P chart to monitor accident frequency would be the most appropriate choice. The sample size for the P chart will be the number of days in the sample. If the data are reported monthly, the sample size for January would be 31 calendar days and for February would be 28 days (or 29 during a leap year). Figure 8.6 is the resulting P chart for accidents by month.

For accident severity, the PIT chose to use an X-bar and R chart to monitor the process. Severity is a continuous variable and is measured by the company in terms

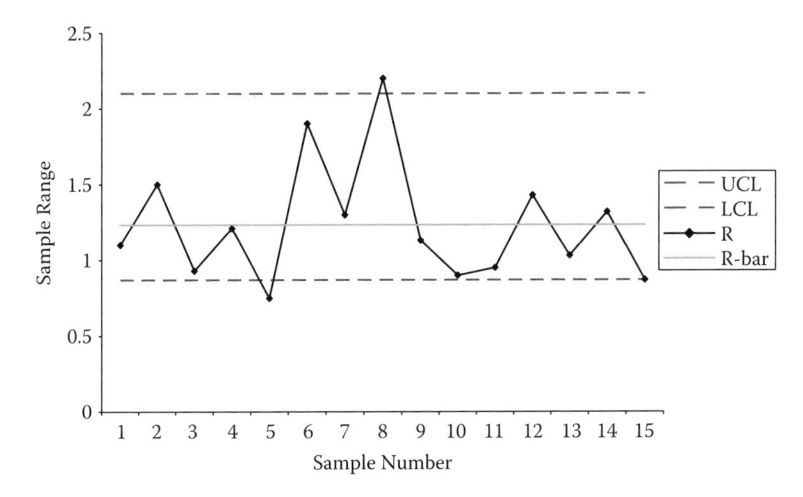

FIGURE 8.8 R chart for accident severity.

of dollars. Based on historical data, the team discovered that the distribution of accident severity did not follow a normal distribution but rather an exponential distribution where there are many accidents with low severity (only small dollar amounts are associated with these types of accidents) and a few extremely severe accidents (associated with large dollar amounts). To correct for this condition, a log transformation was used to normalize the data. The transformed severity data were then plotted as X-bar and R charts as shown in Figures 8.7 and 8.8, respectively.

9 Relationships

9.1 DETERMINING RELATIONSHIPS BETWEEN DATA SETS

On the surface, the word *relationship* seems simple enough to define; however, there is a considerable difference between people relationships and data relationships. At the outset, relationships between people do not exist; they must be developed, and over time they are known to vary as a result of one or more controllable or uncontrollable variables. In his classic 1986 text, *Introduction to Quality Engineering*, Dr. Genichi Taguchi notes that these uncontrollable variables are also known as *noise* variables.

This chapter addresses the subject of relationships between various sets of data. A data set is simply a collection of quantitative values under a common heading; for example, one data set being considered might be the cost to build each home in a given community, whereas another data set might be the number of working days used to construct each of those homes. Figure 9.1 shows how these two data sets compare with one another.

When two or more data sets are being considered, the question often arises as to whether the data sets are related. Furthermore, if a relationship does exist, is it linear or nonlinear (also known as curvilinear)—and if it is nonlinear, is it polynomial, logarithmic, or exponential?

For example, there may be some uncertainty as to the presence or absence of a relationship between the cost of bond exoneration and the number of months until the bond has been exonerated. To determine the presence or absence or such a relationship, take 10 or so x,y data sets where x (time, the independent variable) equals the number of months until a particular bond was exonerated and y (cost, the dependent variable) equals the value of the bond at the time of exoneration. Figure 9.2 is a scatter diagram that demonstrates this concept by including a "best fit" straight line relationship among the multiple bivariate data sets.

9.2 LINEAR RELATIONSHIPS

As discussed in the previous paragraph, a straight line or linear relationship can exist between a data set of x,y data points. Another example of a linear relationship is x (an independent variable), the number of hours worked per week, and y (a dependent variable), the gross take-home pay for a construction superintendent. Figure 9.3 portrays a linear relationship.

The concept of linear relationship suggests that two variables are directly proportional to each other; that is, increasing the independent variable by $X\%$ causes the dependent variable to also increase by $X\%$, and vice versa.

Linear relationships are often the first approximation used to describe any relationship, even though there is no unique way to define what a linear relationship is

Home ID	Cost to Build	Workdays to Build
15-01	$150,000	100
15-02	$165,000	103
15-03	$185,000	110
15-04	$145,000	102
15-05	$175,000	110
etc.		

FIGURE 9.1 Data sets.

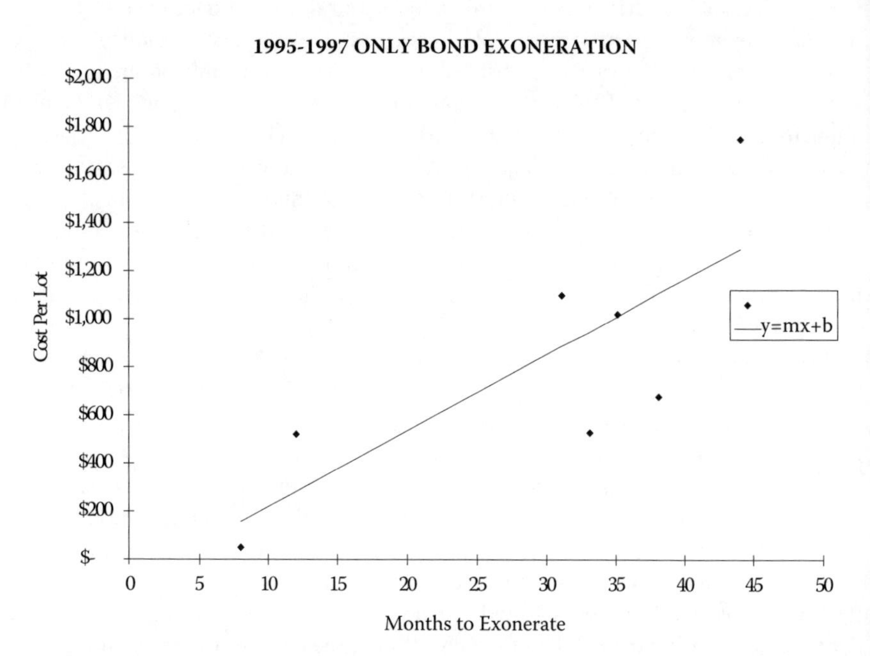

FIGURE 9.2

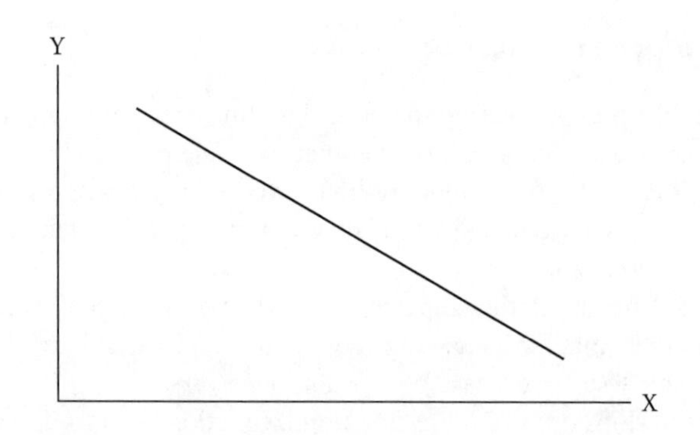

FIGURE 9.3 A linear relationship. This figure portrays a negative relationship, i.e., as "X" (the independent variable) grows larger, "Y" (the dependent variable) becomes smaller.

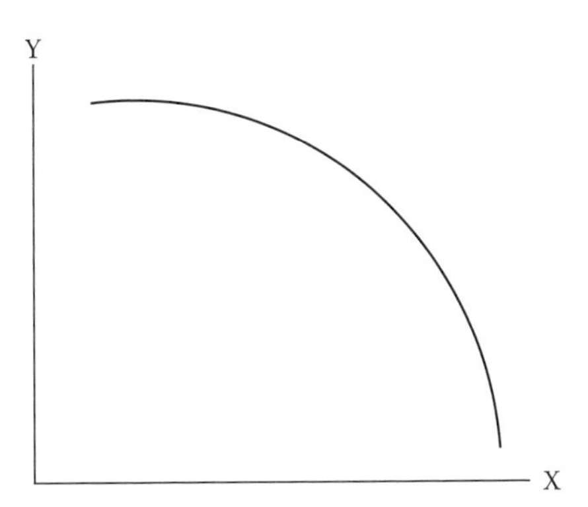

FIGURE 9.4 A nonlinear relationship. This figure portrays a nonlinear or curvilinear relationship, i.e., when the values of "X" are small (near the origin), the associated values of "Y" change quite slowly; however, when the values of "X" are much greater (relatively speaking), the associated values of "Y" change (in this case, drop) rather quickly.

in terms of the underlying nature of the quantities. For example, a linear relationship between the footprint area of a house (measured in square feet) and the cost to build a house (expressed in dollars) is different from a linear relationship between the footprint area of a house and the cycle time to build a house (counted in either working or calendar days). Both relationships make sense and are linear relationships.

9.3 NONLINEAR RELATIONSHIPS

As noted in the previous section, linear relationships are defined by straight lines. However, not all relationships are linear and these are referred to either as nonlinear or curvilinear.

In general, nonlinear relationships are any relationships that are not linear. What is important in considering nonlinear relationships is that a wider range of possible dependencies is allowed. When there is very little information to determine what the relationship is, assuming a linear relationship is simplest; thus, by Occam's razor (which advises simplicity in scientific theories), is a reasonable starting point. However, additional information could reveal the need to use a nonlinear relationship. Figure 9.4 portrays a nonlinear relationship.

Some examples of nonlinear relationship are as follows:

- Increased investment value resulting from compounded interest
- Estimated airflow rates in air-handling units from actuator control signals
- Reduced cost per unit as the number of units purchased increases

9.4 OTHER RELATIONSHIPS

As noted earlier in Section 9.2, there are a number of other relationships in addition to linear and nonlinear. Included among these are polynomial, semi-logarithmic, and exponential. Figures 9.5a, 9.5b, and 9.5c provide visual illustrations of these data relationships.

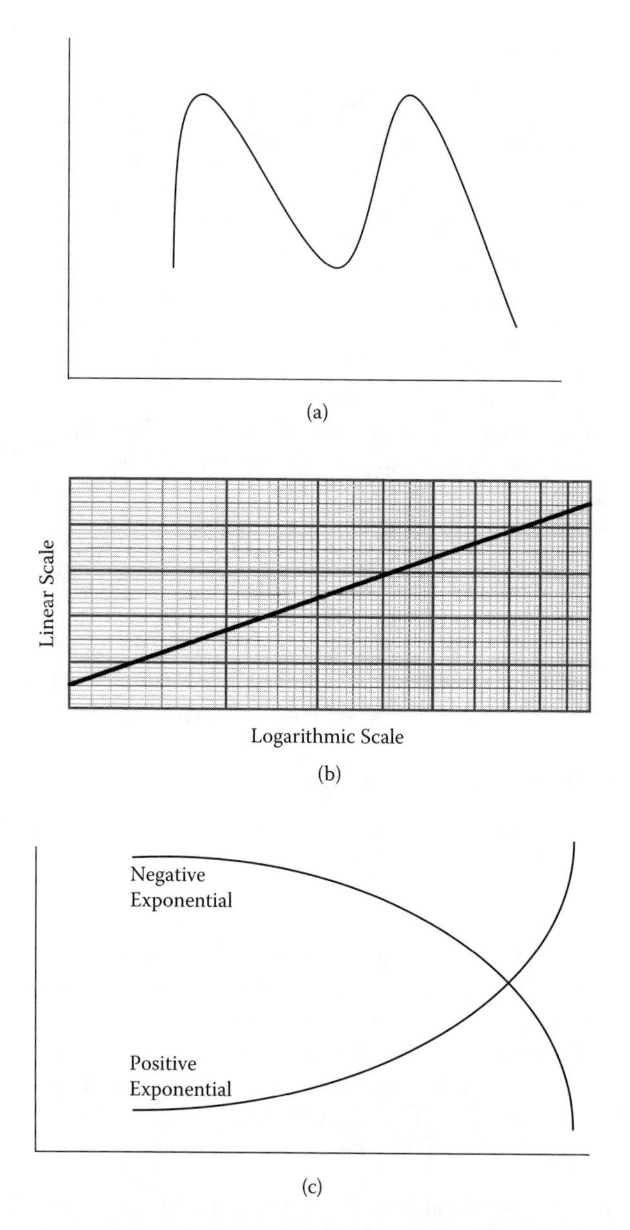

FIGURE 9.5 Data relationships: (a) polynomial, (b) semi-logarithmic, and (c) exponential.

9.5 PREDICTING LINEAR RELATIONSHIPS

Linear relationships are expressed using the following equation (e.g., cycle time as a function of home size in square feet):

$$Y = mX + b$$

where

Y is the value of a dependent variable (e.g., predicted cycle time in days).
m is the slope of the linear relationship (the straight line) = 0.024.
X is the value of an independent variable (e.g., home area in square feet).
b is the value of the Y-intercept (the value of Y when $X = 0$) = 29.

Here's an example. A builder is introducing a new series of homes and needs to estimate how long it will take to build each home for its financial planning. It's been determined from several previous residential construction projects that m is 0.024 and b is 29. The slope and intercept values were obtained using regression analysis of home sizes versus their corresponding cycle times. Note: When you're using most software regression programs, you'll find that the symbols for the slope and the y-intercept vary from those used in this book.

The division manager wants to know what Y (the predicted cycle time) will be if X (the square footage) is 2,450. The answer is 88 days. This was determined using the above equation. Starting out with the equation, we substitute in what is already known and then solve for the unknown value as follows:

$$Y = mX + b$$
$$Y = (0.024)\,(2{,}450) + 29$$
$$Y = 58.8 + 29$$
$$Y = 87.8 \approx 88 \text{ days}$$

Alternatively, suppose the division manager indicates that the new product needs to have an average cycle time Y of 75 days. He wants to know what X (square footage) must be. Once again we start out with the above equation, but this time we substitute in the values of Y, m, and b and solve for X:

$$Y = mX + b$$
$$75 = (0.024)\,X + 29$$
$$75 - 29 = 0.024\,X$$
$$46 = 0.024\,X$$
$$0.024\,X = 46$$
$$X = 46/0.024$$
$$X = 1{,}916.67 \approx 1{,}917 \text{ square feet}$$

9.6 EXAMPLE

Some time ago, the senior author in his capacity as a consulting statistician, was requested by a home builder client to demonstrate how the concept of data relationships could be applied in a way that would assist the builder in relating the quality of the work performed by the builder's employees to the employee's performance evaluation. Figure 9.6A presents the Performance Assessment Scatter Technique (PAST), which graphically examines employee productivity versus product quality.

Part a of Figure 9.6A presents a PAST template with W, the measure of productivity expressed in units per time period on the horizontal axis, and X, the measure of

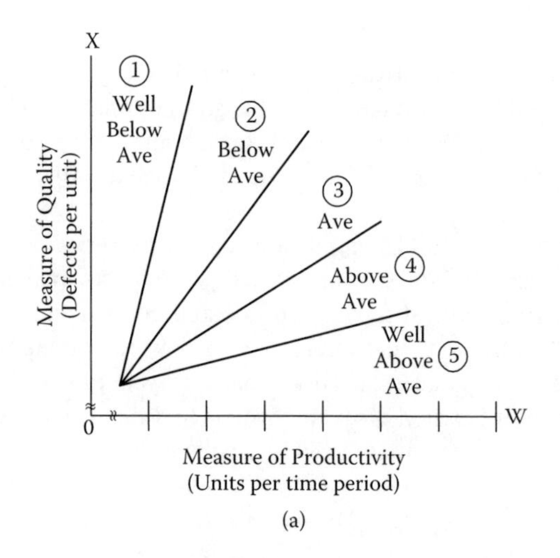

(a)

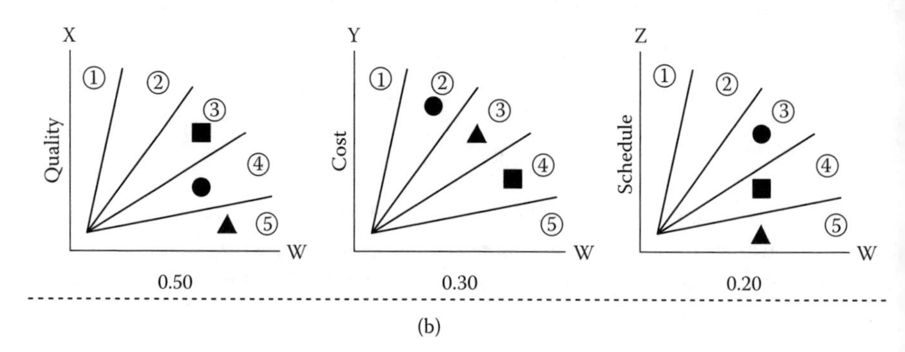

(b)

		Sum	Rank
●	= (.5)(4) + (.3)(2) + (.2)(3) =	3.2	3
■	= (.5)(3) + (.3)(4) + (.2)(4) =	3.5	2
▲	= (.5)(5) + (.3)(3) + (.2)(5) =	4.4	1

(c)

FIGURE 9.6A Performance Assessment Scatter Technique (PAST).

quality expressed in defects per unit on the vertical axis. The L-shaped matrix has been stratified into five employee categories:

1. Well below average: Very high number of defects and very low number of units produced per time period.
2. Below average: High number of defects and low number of units produced per time period.

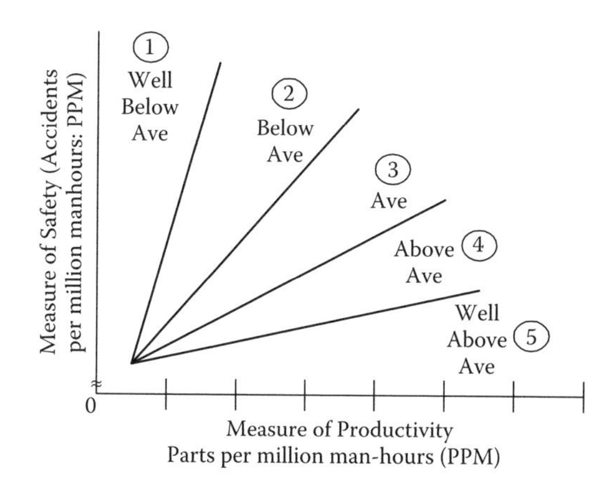

FIGURE 9.7 PAST: Safety performance assessment.

3. Average: Expected number of defects and expected number of units produced per time period.
4. Above average: Low number of defects and high number of units produced per time period.
5. Well above average: Very low number of defects and very high number of defects produced per time unit.

Part b of Figure 9.6A reveals that the PAST approach is not limited to just one aspect of an employee's performance. In addition to the home builder's concern for quality identified in Part a, the PAST is quite capable of accommodating other builders' concerns such as cost and schedule. The example offered in Part b demonstrates one builder's approach, which weighs quality at 50% of the employee performance evaluation, cost at 30%, and schedule at 20%. As might be expected, the concerns as well as their weights can be varied according to the wishes of the home builder.

Part c of Figure 9.6A presents three examples using geometric symbols: one employee is noted using a circle, another with a square, and the third with a triangle. The employee identified with a square had a measure of productivity and a measure of quality that placed her in the *Average* (three points) category for quality (Q), which is worth 50% of the overall evaluation [$Q = (0.5)(3) = 1.5$]. In the cost (C) category, which is worth 30% of the overall evaluation, the lady identified by the square had a measure of productivity and a measure of cost that placed her in the *Above Average* (four points) category [$C = (0.3)(4) = 1.2$]. Finally, this same employee achieved an *Above Average* (four points) in the schedule (S) category, which is worth 20% of the overall evaluation [$S = (0.2)(4) = 0.8$]. Overall, considering quality, cost, and schedule, the employee realized an overall evaluation of 3.5 [$O = Q + C + S = 1.5 + 1.2 + 0.8 = 3.5$].

Since this approach to evaluating employees' performance uses a scale from 1 (lowest) to 5 (highest), the highest score an employee can receive is a 5 and the lowest is a 1. To rank order all the employees, each employee is ranked according to his or her score from the greatest overall evaluation to the least. The scores of three employees in this example had overall evaluations of 4.4 (triangle), 3.5 (square), and 3.2 (circle). Thus, the triangle is ranked as number one, the square as number two, and the circle as number three.

Remember: this was an example using only three areas of concern. Each home builder is encouraged to use whatever areas of concern are most important to its company. Some builders may wish to add areas of concern such as safety, customer satisfaction, or others. As long as the area can be evaluated using a quantitative measure, the list of areas to be considered is virtually unlimited. Figure 9.7 demonstrates how easily this can be accomplished using a safety-oriented example.

10 Sampling and Randomization

10.1 BACKGROUND

In your effort to collect data, no matter what the subject or discipline, you will find that in a vast majority of situations it is either imprudent if not impossible to collect a census of the entire data set. A census is when every eligible person or product has corresponding data. If the objective was to survey your customers about their satisfaction with your products and/or services, a census would mean every single customer would have to be surveyed. If the objective is quality assurance it would mean that each and every home is inspected for every possible defect. The cost of such measures can become staggering to say the least. Fortunately, there is a way to reduce the cost and effort required without substantially sacrificing the quality of the data gathered. The technique is called sampling. This chapter will discuss the strengths and weaknesses of some common sampling procedures being applied throughout business and industry, and that includes residential construction industry.

10.2 CONVENIENCE SAMPLING

Suppose you were interested in determining how your trade contractors would respond to a proposed change in certain payment procedures. You decide to take a sample by asking opinions of subcontractors that come to a regular meeting of your trades on a given Tuesday afternoon between one and three o'clock. This method would be fast and economical, but your sample would not reflect the opinions of the entire population of trade contractors. You have systematically excluded anyone who was absent that day or had something else to do during that time.

Changes based on nonrepresentative information can often be incorrect and biased. If you came up with new payment procedures based only on your convenience sample, the changes could well turn out to be a big mistake. They might, for example, be stiffly resisted by all trades whose workloads were too heavy to permit them to attend that particular meeting and that were therefore systematically excluded from your convenience sample. If you find yourself in a situation when time and financial constraints necessitate convenience sampling, be aware of the limitations of this method.

10.3 JUDGMENT SAMPLING

Returning to the example of changing payment procedures, suppose you decide you'd like to poll approximately 10 trades. There are 10 people whose opinion you respect, so you choose these specific people as the sample. There's a good chance that you respect the opinion of these 10 people because their opinions are similar to yours. If so, you could easily be misled into thinking that your opinion is far more widely shared than it actually is. When using a judgment sample, keep in mind that samples selected on a judgmental basis can sometimes reflect the biases of the judge as much or more than they represent the actual characteristics of the population.

10.4 TYPES OF RANDOM SAMPLING

The essence of random sampling is to set up a process that ensures every element in the population has an equal chance of being selected for the sample. A few of the most common ways to take a random sample are drawing slips of paper from a bowl, using random number tables, and using computer-based random selection procedures. You should choose the method that is most convenient to you based on available resources and the size of the population.

10.4.1 Drawing Slips of Paper from a Bowl

This is a simple method of taking a random sample when the population is not large. Assign a number to each element in the population. Write these same numbers on slips of paper, and put the slips in a bowl. Draw out the slips one at a time, and record each number. The elements corresponding to the numbers you draw become the sample.

Using the payment procedure example again, let's suppose you decide to take the sample randomly. Rather than assigning each trade contractor a number, you simply write each trade's name on a slip of paper. Put all the slips in a bowl, and draw out as many slips as you need for the sample. The names on the slips are your sample. With this method, all the trades have an equal chance of being selected, so your results should be fairly representative of the entire population.

10.4.2 Random Number Tables

A random number table is a table of digits that has been generated by some random process—usually through a computer program. Extensive tables of random numbers are also available in most basic statistics textbooks.

The basic procedure is to sequentially number each element in the population, beginning with 1. Then, randomly select a place to start in the random number table by closing your eyes and touching a point on the page. Sequentially choose as many random numbers from that point as are needed for your sample size, skipping over any duplicate numbers. You can select the numbers by reading up, down, across, and so on, as long as you take them in sequence. The random numbers you select correspond to the numbered elements in the population that you choose as the sample.

Figure 10.1 is a section of a random number table. Even though there are six digits in each random number, you need to use only as many digits in each random number as there are numbers of digits in the maximum population value. It is safe to ignore random numbers that are higher than the maximum population value.

If the number of elements in the population is a power of 10 (e.g., 100, 1,000, 10,000), use only as many digits in each random number as the number of digits in

831770	738428	912923	253596	115656	532319	186532	800513
106919	511567	511782	632383	412057	150937	812164	775564
155582	423547	854377	277846	431445	114748	194197	448768
971218	937174	520230	975625	880153	330058	855701	712527
427076	811130	289554	353140	397919	947619	386436	236258
821487	432208	592848	338211	466129	715198	617306	902459
118066	293076	951104	851004	215503	762617	225915	371594
941214	868582	370991	713715	647617	138919	799735	107429
584699	573814	666010	723165	201409	969958	765175	300228
961314	625647	657139	915552	163505	569481	393433	834508
432469	589550	948747	984827	168356	687256	474025	304994
942985	906480	778231	668987	535590	998210	980357	319394
765697	132931	130655	150478	716338	803957	951713	171415
107310	639260	779851	881985	450718	839850	887060	153944
752588	639182	503215	937612	257260	803486	756939	380933
549739	737851	474813	687807	487101	262263	558090	900105
525141	691378	843256	688329	607268	342199	536094	972237
203194	507603	786222	939711	259697	692827	193751	732473
492898	823218	128032	443300	219338	227604	337658	644708
128641	370879	692183	202399	481119	671042	282188	275986
335765	816310	775249	224692	912674	977819	654203	853542
856823	559955	644154	526568	145972	827543	385531	637381
119059	111615	887683	137180	661762	164882	425360	142408
946137	327673	528709	411503	656827	541401	666024	118013
650330	961509	183109	834048	985107	764467	254755	426410
386746	819996	340268	300768	999197	714195	534314	496339
467508	266398	393632	298169	129223	404962	913092	140348
324334	218418	299380	809097	782553	367448	669724	375039

FIGURE 10.1 Random number table.

the maximum population value minus 1. Done this way, you won't have to skip any random numbers.

To illustrate the use of random number tables, consider the payment procedure example, in which you want to obtain opinions from a sample of 10 trade contractors (i.e., the owners or general managers). Let's say that in this case there are 100 trades, so each trade is assigned a number from 1 to 100.

Suppose the number you land on when you touch the random number table is the right-hand side of 937612 (look for the arrow about halfway down the fourth column from the left). Record the last two digits of each sequential random number from that point. (The population contains 100 elements, so you'll use two consecutive digits in each random number, letting "00" stand for 100). Reading down the table, the random numbers you find and the digits you record are as found in Figure 10.2.

Now find the trade's name in the numbered population that corresponds to each of the numbers you recorded. This would be the 12th name, the 7th name, the 29th name, and so on. These 10 names constitute your random sample.

Random Number	You Record
937612	12
687807	7
688329	29
939711	11
443300	100
202399	99
224962	62
526568	68
137180	80
411503	3

FIGURE 10.2 Recorded digits.

10.4.3 Computer-Based Random Selection

Since random number tables are often generated through a computer program, it is possible to completely computerize the process of taking a random sample. Many computer programs are available to generate the random numbers needed for a desired sample size. Then all you have to do is find the corresponding sample elements in the population you have numbered. Or, if a list of all the population elements is already stored in your computer, it can be programmed to perform the entire random sampling process for you so that the output you get is a listing of the specific elements included in your random sample. Check with your information technology (IT) department about how this is done.

If the characteristic of the population you are studying is the type of data stored in the computer, you can take the process even further by having the data actually collected and analyzed by the computer. For example, suppose you wanted to determine the average salary for all 300 employees in your company. However, rather than using data for all 300 employees, you decide to estimate the average salary based on a random sample. All the data you need is stored in the computer, because the distribution of payroll is a computerized process. You could use a program that assigns a number to each employee, generates the random numbers needed for a sample, locates the salary data of the corresponding sample of employees, and calculates the average salary based on the randomly selected data points.

10.5 APPLICATIONS OF SAMPLING

Usually you study a population to draw conclusions regarding the average or mean value of a given measurement in the population or the percentage of the members of a population that have a certain attribute or characteristic. To obtain the *actual* value, you would have to examine every element. When a sample is used as the base for drawing these conclusions, an *estimate* is made of either the *population mean* or the *population proportion*. Which of these two measures you estimate depends on the nature of the question.

10.5.1 Population Mean

Questions concerning the population mean are those about the average value of a particular measurement in the population as a whole. The data you collect are numerical. The following are examples of questions that involve estimating a population mean:

- What is the average number of incorrect window flashing installations per 100 windows in the homes currently in construction?
- What is the average number of your home sites free of any Occupational Safety & Health Administration (OSHA) violations per 100 home sites during the second quarter of 2009?
- What is the average number of trade contractors' invoices per 100 invoices received that are not paid within the period specified by the terms of your performance contract?

Each of these questions can be answered by finding the average value of the sample data. The average value of the sample is then used to estimate the mean value for the entire population.

10.5.2 Population Proportion

Questions about the population proportion concern what percentage of the population has a certain attribute or falls into a particular category. The data you collect are categorical, or dichotomous (meaning either one or another of two choices, e.g., male or female) in nature. Each element of the population either has the attribute you are studying or does not. The following are examples of questions that involve estimating a population proportion:

- What percentage of homes currently in construction are zero defects in window flashing installations?
- What percentage of your home sites is free of any OSHA violations during the second quarter of 2009?
- What percentage of your trade contractors' invoices have not been paid within the period specified by the terms of your performance contract?

Each of these questions can be answered by finding the percentage of the sample that has the particular attribute being studied. The percentage value of the sample is then used to estimate the percentage of the entire population having the same attribute.

10.6 EXAMPLE

Builder A has recently developed a quality assurance program that includes inspecting homes for defects throughout the construction process. At first, the plan was to inspect every home, but the builder quickly found that given the number of homes it builds each year, it was impractical from a resource standpoint to do so. Therefore, it decided to inspect a sample of the homes built each month and use the resulting sample data to generalize to the greater population.

If Builder A specialized in building one type of product all of a similar size and complexity, a *simple* random sample would be adequate. However, Builder A has two product lines that vary in complexity and size and, as a result, must develop a *stratified* random sampling methodology to accurately estimate how well it is doing. It was decided to create a stratified random sampling plan that would give each product series the same proportional weight in the monthly sample as existed in the overall percentage of production. Thus, if the larger products accounted for 20% of the total homes built by Builder A, they must account for about 20% of the monthly sample within each randomly selected lot.

11 Sample Size Determination

11.1 SAMPLING

Sampling is the selection of a set of elements from a population of products or services. Sampling is frequently used because population data are often impossible, impractical, or too costly to collect. When this is the case, a sample is used to draw conclusions or to make decisions about the population from which the sample is drawn.

When used in conjunction with randomization, samples provide virtually identical characteristics relative to those of the population from which the sample was drawn. Users of sampling are cautioned, however, that there are three categories of sampling error: (1) bias (lack of accuracy), (2) dispersion (lack of precision), and (3) nonreproducibility (lack of consistency). These are easily accounted for by knowledgeable practitioners. A pictorial representation of the types of sampling error is presented in Figure 11.1.

11.2 SAMPLING ERROR

Determinations of sample sizes for specific situations are readily obtained through the selection and application of the appropriate mathematical equation. To determine the minimum sample size, it is necessary to specify the following:

1. If the data are continuous (variable) or discrete (attribute)
2. If the population is finite or infinite
3. What confidence level is desired or specified
4. The magnitude of the maximum allowable error (due to bias, dispersion, or nonreproducibility)
5. The likelihood of occurrence of a specific event

Until you have determined the necessary size of the sample, you cannot really begin to collect the data. There is no simple answer to the question of how large a sample should be. It depends on several interrelated factors. You need to balance the costs of increasing the sample size by a given amount against the benefit you expect to gain in the additional accuracy of the results.

In general, the larger the sample, the more closely your sample results will correspond to the population as a whole. This is true whether you are estimating a population mean or a population proportion. The trick is to determine the sample size that

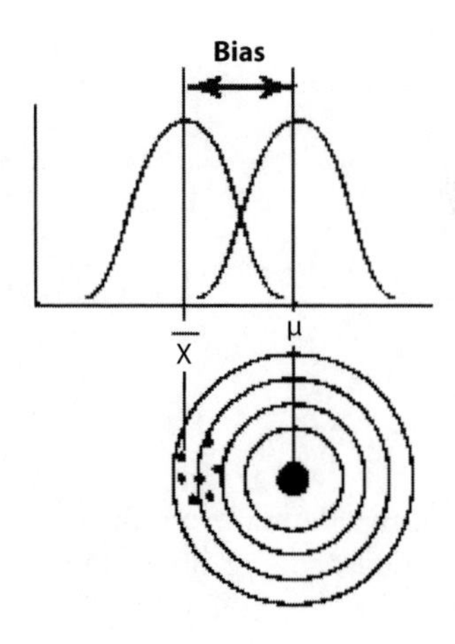

Bias

- Occurs when $\bar{x} \neq \mu$
  (Sample mean does not equal population mean)
- Results from poor instrument calibration

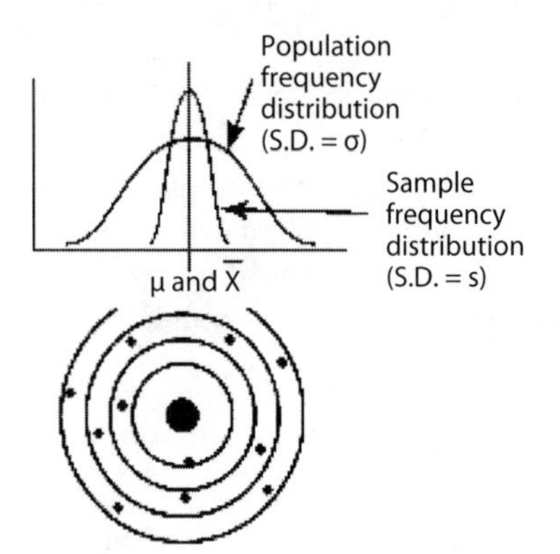

Dispersion

- Occurs when $s > 0$
  (Sample standard deviation does not equal zero)
- Results from improper instrument use

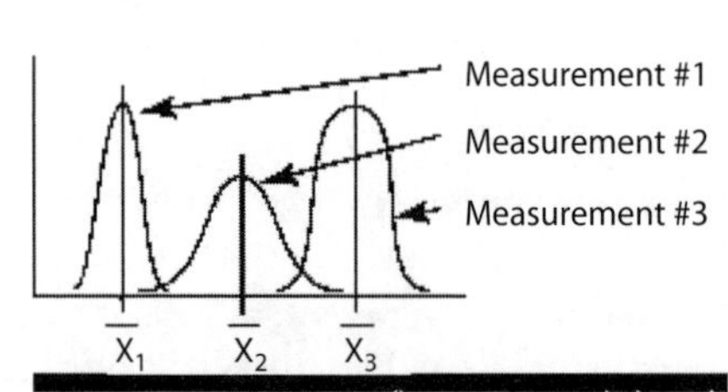

Nonreproducibility

- Occurs when multiple measurements generate different readings
- Results from:
  - Operator or inspector error
  - Instrument calibration error
  - Environmental variables

FIGURE 11.1 Types of sampling errors.

optimizes the balance between the costs of collecting data and the accuracy needed to make a given decision.

The costs associated with a given sample size are dependent on the constraints discussed previously. The accuracy associated with a particular sample size depends primarily on three main factors used to calculate the sample size:

- Population size
- Precision
- Confidence level

11.2.1 Population Size

The required sample size depends to some degree on whether the total number of elements in the population is known, that is, finite. The number of employees in your company is a finite population—you can count them and reach a final number.

However, many populations are assumed to be infinite, that is, too large to count. For example, the elements produced by an ongoing process may be thought of as continuing indefinitely. For example, a computer system that is active 24 hours a day, 365 days a year may be considered to have an infinite population in terms of processing time.

There are different formulas to determine the appropriate sample size for estimating population proportions and population means based on whether you treat the population as finite or infinite. The formulas used to calculate a sample size when you are assuming an infinite population that is actually finite will always result in a sample size that is larger than is actually needed to achieve a given level of accuracy.

If you have a finite population, it is recommended that you use the formula that includes the actual population size so that you do not collect more data than are really needed. If you know the population is finite but you are unsure of the exact size, try to arrive at a reasonable ballpark estimate, and use the finite population formula.

11.2.2 Precision

Precision concerns the following question: How far off the exact population value can your sample estimate be and still be close enough for your purposes; that is, how precise does your estimate need to be? Precision can also be defined as the maximum allowable error (MAE), or the maximum error or deviation from the true population value you are willing to accept.

The precision level specifies the distance (or interval) of values above and below your sample value within which you can expect the true population value to fall. The precision level required for a specific situation is a decision you must make. In setting the precision level, you are specifying an acceptable range of error for your sample estimate.

Do you need to be precise enough so the true population proportion will not differ from the sample percentage by more than 10%? Or do you need to be more precise so the true population proportion can be expected to be not more than 1% higher or lower than the sample value? This decision is based on how far off you think the estimate can be without causing problems when you make a decision based on the sample results.

11.2.3 Confidence Level

The final factor to consider in determining the sample size is this: How confident do you need to be that your sample estimate won't be off by more than the amount you have decided is good enough? How predictable do you want your estimate to be?

The appropriate confidence level depends on the risk you are willing to take in using the sample results to generalize to the entire population. Are you willing to accept a probability of 75%—or three of four chances—that the actual population

Confidence Level	Z-Value
99.7%	3.000
99%	2.576
95%	1.960
90%	1.645
85%	1.440
80%	1.282
75%	1.150

FIGURE 11.2 Z-values.

value will be within your specified precision level (MAE)? Or do you want a higher probability, say 90% (9 of 10 chances), that your estimate will not differ from the true population value by more than the amount you specified as acceptable? This decision is based on the practical consequences of being in error by more than the amount you specified in the precision level.

For purposes of determining the sample size, the confidence level is expressed by a coefficient called a Z-value. The theory behind Z-values is not discussed here, since for most applications you need only to designate the confidence level you want and pick the corresponding Z-value to use in the formula. Figure 11.2 shows the Z-values that correspond to the most commonly used confidence levels.

The formulas used to determine the sample size are different depending on whether the question concerns estimating a population proportion or estimating a population mean. These formulas are discussed separately.

11.3 POPULATION PROPORTION

The sample size needed to estimate a population proportion is determined primarily by the confidence level and the precision level you want to achieve. The *confidence level* is the probability that the sample estimate will not differ from the true population proportion by more than the amount specified by your precision level. The *precision level* is the MAE in your sample results.

The variables used to calculate the sample size to estimate a population proportion are as follows:

n = size of sample needed
N = size of the population (finite)
Z = the Z-value from the figure that corresponds to the specified confidence level
d = the specified precision level (expressed as a decimal fraction, i.e., 10% = 0.10)
p = the true population percentage (expressed as a decimal fraction, i.e., 50% = 0.50)
$q = 1 - p$

The variable p represents the true proportion of the population that has the particular attribute or falls into the particular category you are interested in estimating. And, since the remaining elements would not fall into that category, these represent q, or $1 - p$.

An exact value for p is usually impossible to determine, since it is the value you are interested in estimating, but you can usually make an educated guess. If you are estimating a population proportion for a situation that you have studied before, you may have some prior data on which to base an educated guess. Whatever the sample percentage was the first time can be used as the value for p.

If you have no prior data, you can usually still make an educated estimate for p, but there is some risk involved. The further the percentage you designate for p deviates from 50%, the smaller the calculated sample size will be. So, if your estimate of p is quite inaccurate, your sample size may in fact yield actual precision and confidence levels that are very different from what you had intended to achieve.

If you have no basis at all for making an educated guess about what the true population proportion is likely to be, the most conservative estimate you can make is that there is a 50:50 chance that the element either has the characteristic or does not. You will always be safe by using $p = 0.50$ and $q = 0.50$, although in some cases this may yield a sample size that is substantially larger than is actually needed to achieve the desired confidence and precision levels.

To calculate the exact sample size needed, taking into account a finite population size (N) and some estimate for the true population percentage (p), use the following formula:

$$n = \frac{NZ^2pq}{Nd^2 + Z^2pq}$$

If you assume an infinite population and the most conservative estimate (i.e., $p = 0.50$) for the true population percentage, use the following formula:

$$n = \left[\frac{0.5z}{d}\right]^2$$

This second formula assumes a true population percentage of 50% and an infinite population. It always yields a sample size that is somewhat larger than the sample size calculated according to the first formula, which takes into account the actual (or estimated) size of a finite population.

11.4 EXAMPLE: POPULATION PROPORTION

Suppose you are responsible for the development and deployment of a comprehensive sampling plan for use by your company. Specifically, the plan should be designed to include each home under construction. This plan should focus on all facets of window installation including window flashing (sill, jamb, and head), window rating, window size, and window assembly.

You are expected to create a plan that requires the full inspection of five windows of each window style in each home selected for inclusion by the random sampling plan. In this situation, there are an average of three window styles in each home and a total of 200 homes in four subdivisions. Thus, the total number of windows (the population) is computed to be 3,000.

$$\text{Total} = \frac{3\,\text{window styles}}{\text{home}} \times 200\,\text{homes} \times \frac{5\,\text{windows}}{\text{window style}}$$

$$= 3,000\,\text{windows}$$

The question you want to answer is, "What percentage of the windows has been installed improperly?" You need to know the population percentage to determine if the percentage is decreasing over time (from month to month).

To inspect all 3,000 windows would be both time-consuming and costly. You decide that sampling a random selection of windows will provide an adequate estimate of the percentage of windows that have been installed improperly.

To determine the appropriate sample size, begin by deciding how far your resulting estimate can be—either high or low—from the true population value without causing any problem in understanding the actual percentage of all the windows that have not been properly installed.

You want to make sure that the true percentage of all 3,000 windows, the population, that are improperly installed won't be more than 10% higher or lower than the percentage of windows in the sample that are improperly installed. You have to set the precision level of your sample not to exceed ±10%. But how sure do you want to be that your sample estimate won't be off by more than 10%? This is an issue of your desired (or required) confidence level.

After discussions with your president and vice president of operations/construction, you decide you want to have 8 of 10 chances, or a probability of 80%, that your sample percentage will be within 10% of the actual population percentage. In making this decision, you have set the confidence level at 80%.

Given that you will accept a ±10% precision level and an 80% confidence level and that you know that the population size (N) is 3,000, you can now compute the required sample size. According to Figure 11.2, the Z-value that corresponds to a confidence level of 80% is 1.28. Since you have no prior data on the true population percentage, you use 0.50 for the p and q values:

$$\text{Required sample size} = \frac{NZ^2pq}{Nd^2 + Z^2pq}$$

$$= \frac{3,000 \times 1.644 \times 1.644 \times 0.50 \times 0.50}{(3,000 \times 0.01 \times 0.01) + (1.644 \times 1.644)(0.50 \times 0.50)}$$

$$n = \frac{1233}{30.411} = 40.5$$

Rounding upward yield

$$n = 41$$

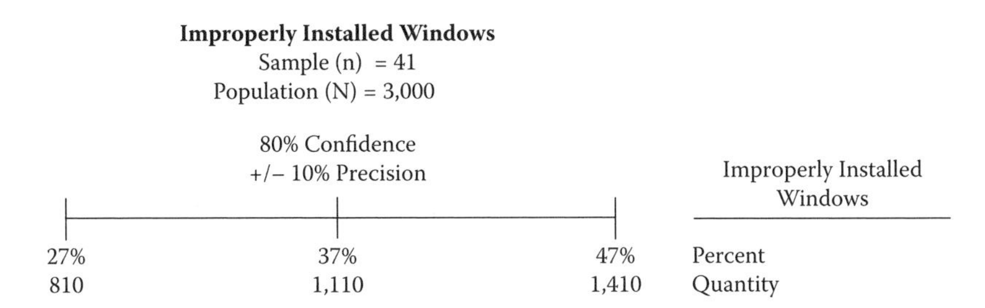

FIGURE 11.3 Results of windows example.

By taking a random sample of 41 of 3,000 windows, you have an 80% chance that the percentage of improperly installed windows in the sample will not deviate from the percentage of the entire population by more than 10%.

After inspecting these 41 windows that were selected at random, you find that 37% of them (15 windows) have been installed improperly. Since the precision level was ±10%, the percentage of the entire population that is improperly installed should be between 27% and 47%. Figure 11.3 indicates that you can be 80% confident that between 810 and 1,410 windows are improperly installed, which is 27% and 47% of 3,000, respectively.

Using this sampling plan was clearly a cost-effective way to determine the effectiveness of the window installation process because the cost and time to inspect 41 windows is minimal, but the cost of inspecting all 3,000 windows (not to mention the cost of processing the results) is excessively high.

Figure 11.4 shows other sample sizes you would need to attain various combinations of confidence and precision for the windows example. If you wanted to increase your chances of not missing the true population percentage, you would need to increase the sample size. For example, if you wanted to be 95% sure you would not miss the true population value by more than 10%, you would need to inspect 94 windows. If you wanted to narrow the interval around the sample percentage where you expect the true population percentage to be, you would also need to inspect more windows. Suppose you wanted to be 80% sure that you would not miss the true value by more than 5%; this translates to ±5% precision and 80% confidence, or 156 windows.

Increasing either the precision or confidence will increase the sample size, but precision is much more costly in terms of the sample size required. By the time

	Confidence Levels			
Precision	**80%**	**90%**	**95%**	**99%**
±10%	41	67	94	158
±5%	156	249	341	544
±1%	1,734	2,079	2,286	2,541

FIGURE 11.4 Sample size needed to estimate the population percentage (population size = 3,000).

	Confidence Levels			
Precision	**80%**	**90%**	**95%**	**99%**
±10%	42	68	97	166
±5%	165	271	385	664
±1%	4,109	6,676	9,604	16,590

FIGURE 11.5 Sample size needed to estimate the population percentage (population assumed *infinite*).

you get to ±1% precision and 99% confidence, you have 2,541 windows in the sample. If the results really need to be this accurate, you may as well inspect all 3,000 windows.

The sample sizes listed in Figure 11.4 were calculated based on a *finite* population of 3,000 windows. For purposes of comparison, the sample sizes needed assuming an *infinite* population are shown in Figure 11.5.

As you see, there is not much difference between the results of the finite versus the infinite population formulas at the lower levels of precision and confidence. As precision increases, however, the differences can become quite large. As this comparison shows, then, if your population size is relatively small or you have some very stringent precision requirements, you would do much better to calculate the sample size via the finite population formula so that you don't collect more data than you actually need. This would be especially true if collecting the data is quite expensive or difficult.

11.5 POPULATION MEAN

As with estimating a population proportion, the sample size needed to estimate the population mean is determined primarily by the confidence level, the precision level, and whether the population size is finite or infinite.

Confidence refers to the probability that your sample average will not differ from the true population mean by more than the amount specified by your precision level. The precision level is the interval of values above and below your sample mean within which you expect the true population mean to occur.

The precision level for estimating a population mean is expressed in standard deviation units. Once the sample mean is determined, you calculate the sample standard deviation to find the actual interval of values that correspond to the specified precision level.

The standard deviation is a measure of the variability or dispersion of all the data values in a given data set or distribution. Computerized statistical packages are available that will calculate the standard deviation of any set of data. Excel is a good example. Many hand calculators are also programmed to perform this function. Formulas for calculating the standard deviation from ungrouped and from grouped data, along with brief descriptions of the procedures, are available in Chapter 5.

It is possible, however, to gain a rough conceptual grasp of the meaning of a standard deviation without actually performing the calculations. One way to do this is to understand how standard deviation units relate to the proportion of data values contained within various intervals above and below the mean in a normal distribution.

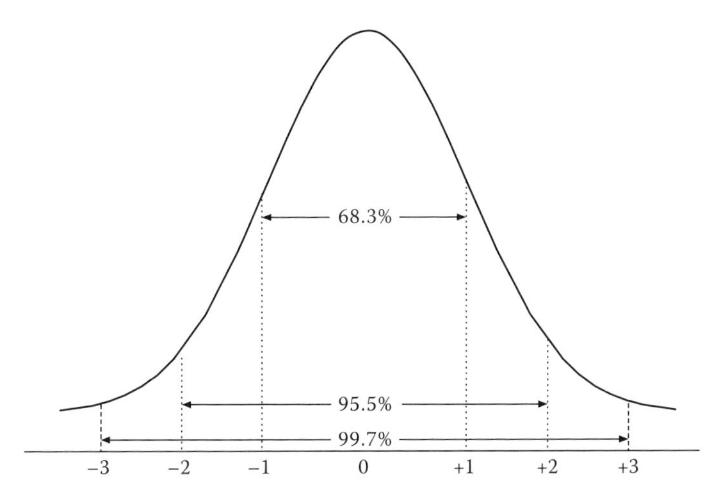

FIGURE 11.6 Percent of scores in a normal distribution contained within various intervals expressed in standard deviation units.

If the distribution of a set of data is a normal distribution, the data values cluster around one average value, called the mean. As the data values get farther away from the mean, they occur less frequently. Approximately the same number of data points occur above and below the mean and the curve is bell-shaped and symmetrical. If a distribution is normal, it is possible to specify precisely what percentage of data falls between any two values, where the values are expressed in standard deviation units. This concept is illustrated in Figure 11.6.

Specifically, in a normal distribution, you can infer that 68.3% of the data will be contained within one standard deviation value above and below the mean (identified as 0 on the graph); 95.5% of the data will be contained within two standard deviations from the mean, and 99.7% of the data will be contained within three standard deviations from the mean.

In a normal distribution, then, you can expect 99.7% of the data to be contained within three standard deviations above and below the mean. Consequently, if you were to divide a normal distribution into six equal parts, each part would represent roughly one standard deviation unit as shown in Figure 11.7.

So, although the approximation is relatively crude (especially when the data set is not extremely large or the distribution is only approximately normal), one way of grasping the meaning of a standard deviation is to realize that it represents approximately 1/6 of the distance between the highest and lowest score in a normal distribution.

Now that we have briefly discussed the meaning of a standard deviation, let's return to estimating a population mean and show how the standard deviation that is calculated from a sample set of data can be used to specify the interval within which the true population mean can be expected to fall.

Suppose, for example, that the average of a set of sample data turned out to be 10 and that the standard deviation of the sample data was 3. If the size for this sample was determined by a confidence level of 95% and a precision level of ±1 standard

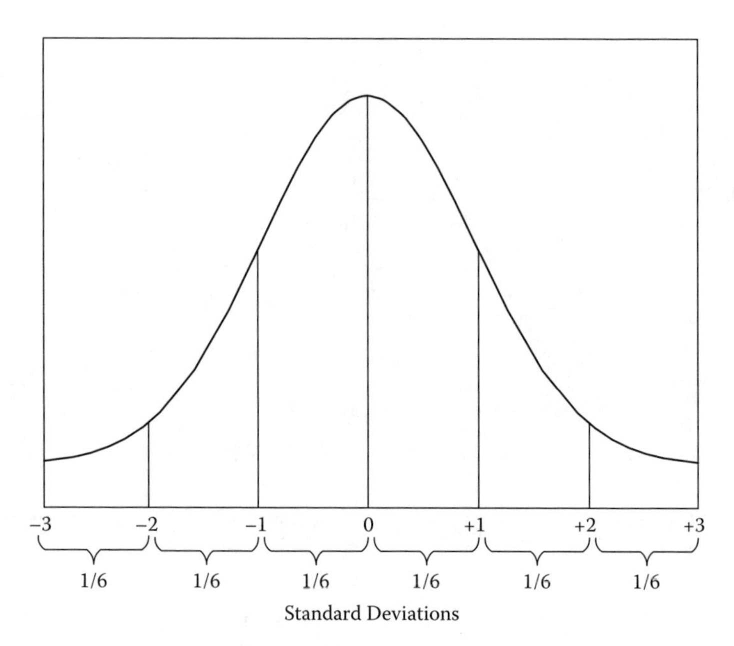

FIGURE 11.7 Standard deviation units under a normal curve.

deviation, you could infer with 95% confidence that the mean value of the population was between 7 and 13 (i.e., 10 – 3 and 10 + 3), as Figure 11.8 indicates. If the sample size had been determined at a confidence level of 95% and a precision level ±1/3 standard deviation, you could infer with 95% confidence that the mean value of the population was between 9 and 11 (i.e., $3 \times 1/3 = 1$; $10 - 1 = 9$; $10 + 1 = 11$) as Figure 11.9 indicates. The higher the precision level, the smaller the range of error in your sample estimate.

Once you have calculated the value of one standard deviation for the sample data, you can specify the interval of values above and below your sample mean within which you expect the true population mean to fall. At a precision level of 1, you expect the population mean to fall within 1 standard deviation value above or below the mean. To be more precise, you would set the precision level to some portion of 1 standard deviation. Figure 11.10 illustrates this concept.

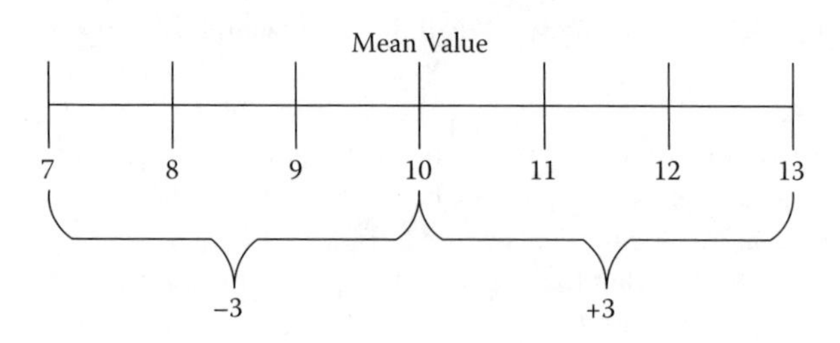

FIGURE 11.8 Precision = ±1 standard deviation.

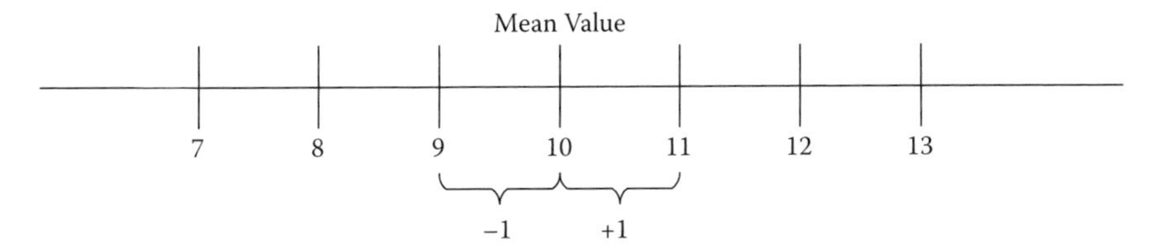

FIGURE 11.9 Precision = ±1/3 standard deviation.

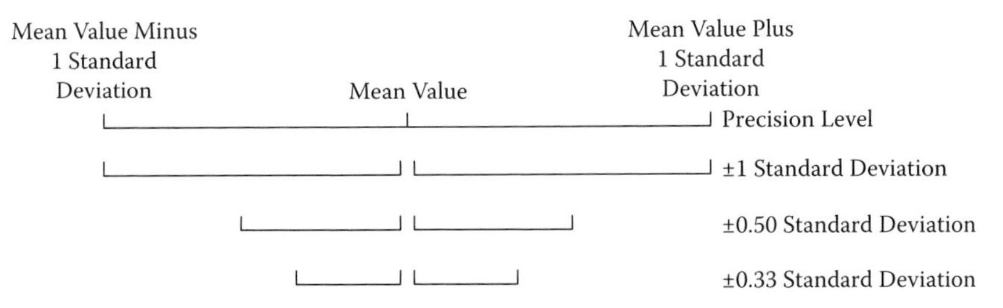

FIGURE 11.10 Precision levels for population mean.

The variables used to calculate the sample size to estimate a population mean are as follows:

n = size of sample needed
N = size of the population (finite)
Z = the Z-value from the figure that corresponds to the specified confidence level
d = the specified precision level (expressed in standard deviation units)

Each Z-value in Figure 11.11 is associated with a particular confidence level and is the same as those used to calculate the sample size to estimate the population proportion.

If you want to calculate the sample size exactly, taking into account a finite population size (N), use the following formula:

$$n = \frac{NZ^2}{Nd^2 + Z^2}$$

Confidence Level	Z-Value
99.7%	3.000
99%	2.576
95%	1.960
90%	1.645
85%	1.440
80%	1.282
75%	1.150

FIGURE 11.11 Z-values.

	Confidence		
Precision	**90%**	**95%**	**99.7%**
$\pm 1.00\sigma$	3	4	9
$\pm 0.50\sigma$	11	16	36
$\pm 0.33\sigma$	25	36	83
$\pm 0.25\sigma$	44	62	144
$\pm 0.20\sigma$	68	97	225
$\pm 0.10\sigma$	271	385	900

FIGURE 11.12 Sample size needed to estimate the population mean (population assumed *infinite*).

If you assume the population is infinite, use the following formula:

$$n = \frac{Z^2}{d^2}$$

The infinite population formula always yields a required sample size that is larger than the sample size calculated according to the finite population formula. The various sample sizes needed to estimate the population mean, assuming an infinite population, are shown in Figure 11.12. These sample sizes can be used with any population.

11.6 EXAMPLE: POPULATION MEAN

Let's suppose you are in charge of testing a fire extinguisher for use in the trucks used by your company's superintendents, customer care personnel, and safety manager.

You intend to conduct the test program from a quality standpoint: How long does a fire extinguisher stay charged? Consequently, you need some data on the average life of this type of fire extinguisher to be able to make a valid statement to the company's owner.

To test the life of a fire extinguisher, you actually have to store it until the indicator dial shows it has run down, so to preserve the population you take a sample. Since the production of this product is an ongoing process, you assume an infinite population.

You need to be very confident about the average life, and the fire extinguishers are not really all that expensive in terms of loss, so you set the confidence level at 99.7% with a precision level of ±0.20 standard deviation. The Z-value in Figure 11.11 that corresponds to a 99.7% reliability level is 3.

The sample size you need is

$$n = \frac{Z^2}{d^2} = \frac{(3.000)^2}{(0.20)^2} = \frac{9}{0.04} = 225$$

You randomly test 225 extinguishers and subject each one to a forced life test. You record the maximum amount of time that each extinguisher keeps a charge. Then, calculate the mean and standard deviation of the extinguisher life in hours

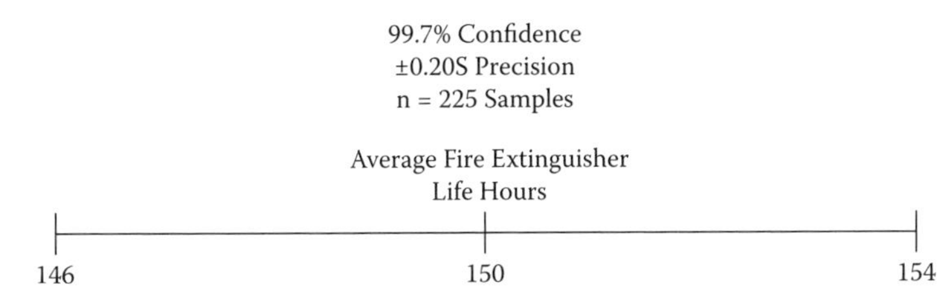

FIGURE 11.13 Results of fire extinguisher example

based on the 225 sample measurements. You find that the mean is 150 hours and that the standard deviation is 20 hours.

Since you used a precision level of ±0.20, or 2/10 of 1 standard deviation, multiply 20×0.20 and find a precision value of ±4 hours.

You can, therefore, infer with 99.7% confidence that the average life of all the fire extinguishers used in your company's trucks will be between 146 and 154 hours. This is the basis you use for your statement to the company's owner. The results of the sample study are shown in Figure 11.13.

12 Software

12.1 INTRODUCTION

In the early 1980s, at the outset of the quality movement in the United States, most companies that were serious about continuous improvement collected and analyzed their data manually. But that was then, and this is now. Since those times computers, which have become considerably faster in crunching data and now possess greater capacity for data storage, have become a regular part of our daily work. Over the last 20 or so years, an analysis that took a computer several hours to calculate is now performed in fractions of a second. This amazing increase in computing power has improved both our productivity and our ability to predict and model future performance. Home construction sequences can be simulated thousands of times with unique duration probabilities assigned to each task to estimate the most likely duration and expected variation of a new process or process change. Complex experimental designs can be run to optimize processes or products for certain attributes or characteristics in what seems like no time at all. The longest part of the analysis process is no longer the computations but rather the setup and interpretation of the data.

There are myriad computer software packages ranging from simple to complex that are available to meet virtually any computational or analytical need that arises. This chapter is a brief overview of some of the most popular continuous improvement software packages as well as their strengths and limitations that are available to home builders.

12.2 EXCEL

One of the most widely available and deceptively powerful tools is Microsoft Excel. With Excel, data tables and their corresponding bar charts, line charts, scatter diagrams, and pivot tables can be quickly created. It contains preloaded formulas that will calculate, for example, sums, averages, and standard deviations. Excel also includes a data analysis add-on that can be installed to perform basic regression, analysis of variance (ANOVA), and t-tests.

The data analysis pack can be found under the tools menu by selecting the add-ins. Once installed, the data analysis pack will always be available under the tools menu for future use. Excel is quite straightforward to use once you get comfortable with it and is the program most likely to be available to you without having to purchase additional software.

However, Excel does have considerable drawbacks when trying to perform complex analysis or analysis of large data sets. The following table provides a list of common functions and their associated formulas to be used in Excel. The (...) notation represents the selection of cells in the Excel workbook that should be used in the function. For example, if you wanted to sum the data in rows 1–12 in column A the formula would be =sum(A1:A12).

Function	Description	Formula
Sum	Adds all numbers in a range of cells	=sum(...)
Average	Calculates arithmetic mean of selected values	=average(...)
Standard deviation	Calculates standard deviation of a sample	=stdev(...)
Correlation	Calculates correlation between two arrays of values	=correl(...)
Median	Calculates median value from a list of numbers	=median(...)
Rank	Returns rank of a number from a list of numbers	=rank(...)

12.3 QI MACRO FOR EXCEL

While Excel has many built-in functions, it does not have some of the most commonly used quality tools and charts built into the standard program. While Excel can be used to create these graphs manually, other software add-ins will allow you to create these charts in a single click. One add-on that is particularly easy to use and useful is QI Macro for Excel. QI Macro provides user-friendly access to many dozens of continuous improvement tool templates and guides such as affinity analysis, bar charts, box and whisker charts, cause and effect (fishbone) diagrams, CUSUM charts, failure modes and effects analysis, force field analysis, frequency distribution curves, histograms, line graphs, multivariate analysis, Pareto analysis, pie chart, process flowchart, scatter analysis, statistical quality control (SQC)/statistical process control (SPC) charts, tree diagrams, and voice of the customer (VOC) matrix. Its versatility and low cost makes this software a popular favorite among home builders.

12.4 ADVANCED ANALYTICAL SOFTWARE

For more advanced analytics, it is necessary to step up to a more powerful statistical platform like Statistical Package for the Social Sciences (SPSS), Minitab, or SAS. All three share many of the more common advanced analytical tools, but they do have some important differences in use and in areas of emphasis.

12.4.1 Minitab

Of the three, Minitab is geared the most toward process improvement and quality tools. It performs particularly well in the area of SQC and SPC statistical control charts and design of experiments (DOE). Like the others, it has both a graphical user interface (GUI) with selectable analysis as well as an option for syntax code entry. For most users the analysis available in the GUI will be more than enough. It is not without weaknesses, however, and some of these weaknesses are admittedly our opinion and it may or may not be shared by others.

Minitab has historically been very limited in the options available to deal with missing data. Generally when dealing with designed experiments and statistical process control, missing data are not much of an issue. The amount of missing data in these cases should be minimal due to the designed nature of the research. However, when performing analysis on recorded data such as consumer data or even financial or safety data collected by the company for its records, the probability of encountering missing data greatly increases.

12.4.2 SPSS

SPSS is a statistical analysis package that is considerably more powerful and flexible than Excel and its various add-ins. It is predominantly used in social science research, which includes customer satisfaction and behavior data. Its has several built-in options for dealing with missing data within the data set and is quite useful in comparing different groups within a data set or filtering the data. Its strengths are in regression analysis, general linear modeling, factor analysis, and cluster analysis. While the GUI incorporates many advanced analytical tools, some analyses like multivariate ANOVA still require syntax or programming to be entered to run.

With the more powerful programs that are fairly easy to use, there is a temptation to run analyses that a novice user may not necessarily completely understand. Because these statistical tests are based on explicit assumptions that if violated jeopardize the results of any analysis, we recommend that an experienced person with proficiency in data analysis be consulted before performing analysis outside the competencies present within the group or company.

SPSS has become a leader in "predictive analytics" technologies through a combination of commitment to innovation and dedication to customers. SPSS customers are found in virtually every industry, including telecommunications, banking, finance, insurance, health care, manufacturing, retail, consumer packaged goods, higher education, government, and market research.

Customers use SPSS predictive analytics software to anticipate change, manage both daily operations and special initiatives more effectively, and realize positive, measurable benefits. By incorporating predictive analytics into their daily operations, they become predictive enterprises and are able to direct and automate decisions to meet business goals and to achieve measurable competitive advantage.

12.4.3 SAS

SAS is another powerful statistical software program that can handle enormous data sets easily and reliably. However, SAS is still heavily reliant on syntax or programming to be entered to perform analyses. It takes some time to understand the programming language, but once understood there are few limitations to the program. The downside of programs such as SAS and SPSS is the cost.

Today's organizations are dealing with diverse issues, a wider range of regulations, and heightened global competition. There has never been a greater need for proactive, evidence-based decisions and agile strategies. With SAS business analytics, your information assets can be transformed into true competitive advantage. By delivering insights that are gleaned from data about customers, suppliers, operations, performance, and more, this software empowers its users to do the following:

- Solve complex business problems
- Manage performance to achieve measurable business objectives
- Drive sustainable growth through innovation
- Anticipate and manage change

12.5 DESIGN EXPERT

Design Expert is a program specializing in DOE. It has the capabilities of creating, for example, factorial designs, fractional factorial designs (e.g., Taguchi orthogonal arrays), and response surface methodology (RSM) designs (Box-Behnken). Design Expert does not require Syntax or programming to be entered by the user to perform analyses. It is both user friendly and reasonably priced.

Design Expert has considerable flexibility, such as the following:

- Ignore a row of data while preserving the numbers
- Add new factors and blocking to existing designs
- Change factors from numerical to categorical and back
- Custom design generators for fractional factorial designs
- Change models from polynomial (RSM) to factorial and back

12.6 QFD SOFTWARE

QFD is a system used to translate the voice of the customer (verbatims) into specific customer requirements (WHATs) to directly related engineering requirements (HOWs). QFD uses cross-functional teams of subject matter experts (SMEs) to assign customer requirements throughout the supplier's organization. QFD is the first phase of the three-phase variability reduction process (VRP). Phase 2 is DOE, and Phase 3 is SPC. The output of QFD becomes the input to DOE; the output of DOE becomes the input to SPC.

QFD uses a collection of interrelated matrices to analytically determine where and how scarce resources should be applied to maximize customer satisfaction in less time and with lower costs.

12.6.1 QFD Capture

QFD Capture (Professional Edition) is an excellent tool for any planning process, from basic to complex. The software has a decision model focus (the roadmap) rather than a single house of quality (HOQ) focus. Having a decision model focus means it is possible to set up and interlink a collection of lists, matrices, and documents to construct a decision-making model for a project. This decision-making model can be high level (a single HOQ) or detailed (many HOQs cascaded from one to another) or something in between.

A QFD Capture model enables a process improvement team (PIT) to collect, analyze, and manage qualitative data to provide an accurate definition of the competition, customer expectations, and the needs of the business. The model will also do the following:

- Help develop a prioritized list of what customers expect to see in products, services, or strategies
- Help translate customer expectations into specifications that designers can understand and act on
- Help develop an ordered list of what steps a company must take to satisfy customer requirements
- Help plan products, services, or strategies with fewer midpoint corrections

12.6.2 QFD Designer

12.6.2.1 Tool for Business Improvements (Compatible with MS Windows 95/2000 or NT)

Realize quicker payback on your policy deployment projects, quality function deployment projects, business planning projects, design failure mode and effects analysis (DFMEA), process failure mode and effects analysis (PFMEA), and best practices analysis with this feature-rich tool. It provides what is needed to create, manage, and advance all your business improvement work in one package. QFD Designer is an easy-to-use, interactive software package for improving quality and customer satisfaction.

QFD Designer provides numerous important features such as the following:

- Work right on chart
- Cut, copy, and paste chart images for easy reporting
- Quick templates
- Lock WHATs/HOWs while you scroll
- Zoom out to easily preview and navigate
- Link to Internet/intranet sites
- Extract subset charts to divide work among the team

Other features include the following:

- Stretch/shrink chart regions
- Unlimited chart size
- Hide/unhide (keeps key data confidential)
- Place custom rooms anywhere

12.6.3 QFD Online

For quality function deployment, QFD Online is online software that allows one to easily create, store, and share the HOQs generated by users.

12.6.3.1 QFD Online's "QFD Builder"

Because of the tremendous usefulness associated with a browser-based house of quality solution, QFD Online plans to develop a Web-based QFD Builder application. They intend to offer this software at no cost (yes, that's what they said) as a service to the QFD user community. It is the company's hope that a Web-based QFD application will facilitate greater collaboration among QFD practitioners as well as further the adoption of the HOQ tool.

12.7 MICROSOFT PROJECT

When it comes to project management software, there are many programs from which to choose. Microsoft Project is one such widely used program. MS Project allows a project manager to account for both the scheduled tasks to be completed as well as the management of available resources. Each task in MS Project has an assigned duration as well as defined predecessor events or tasks that must be completed prior to a given task. With all of the durations and relationships defined, MS Project creates the schedule to minimize the overall duration and cost of the project. As with any program, the output is only as good as the data being entered. If estimations of task durations are not realistic, one cannot expect the actual project duration to match the estimated project duration.

MS Project also has a number of add-in programs available that enhance the capability of the software. One add-in of note is @Risk for MS Project. With this program, one can add probability distributions to each task duration and can run simulations of the project duration to estimate the actual real-world cycle time as well as to estimate cycle time improvements that are achievable by reducing variation throughout the process. Essentially, this program allows one to add variation to each task as it is going to be completed. In one simulation the task may take one day; in another it may take three. These boundaries and the likelihood of occurrence are defined by the user to make the software very good at providing useable and reliable results.

13 Continuous Improvement Tools and Techniques

13.1 INTRODUCTION

The tools and techniques discussed in this chapter and throughout this book can be classified in a variety of ways. One way to look at them is as data transformers. Data transformers have been subclassified by their potential application as being either descriptive tools and techniques or diagnostic tools and techniques. A second way to consider the complete collection of tools and techniques is known as targeted objectives, which have been subclassified as being either analytical or prescriptive. Figure 13.1 presents a graphic summary of these classifications.

Sections 13.2 and 13.3 discuss these classifications and subclassifications and what they mean to home builders. Section 13.4 introduces four tables designed to assist neophyte users in the selection of the most commonly tools and techniques. The final portion, Section 13.5, provides a comprehensive but succinct discussion of all the tools and techniques touched on in the first 12 chapters.

13.2 DATA TRANSFORMERS

13.2.1 Descriptive Tools

Descriptive tools are designed to transform raw data from an unsorted collection of independent and unique values into one or more selected graphics that can help home builders, their trades, and their customers better understand the characteristics of the population from which the raw data were drawn.

13.2.2 Diagnostic Tools

Diagnostic tools are used to transform raw data from an unsorted collection of unique and independent values into statistical estimates of the parameters (characteristics) of the population from which the data were drawn. Just as medical doctors use various chemical and radiological tests to diagnose and better understand their patients' physical problems, so can home builders identify the nature and importance of their construction and administrative problems (i.e., root cause analysis) using diagnostic tools.

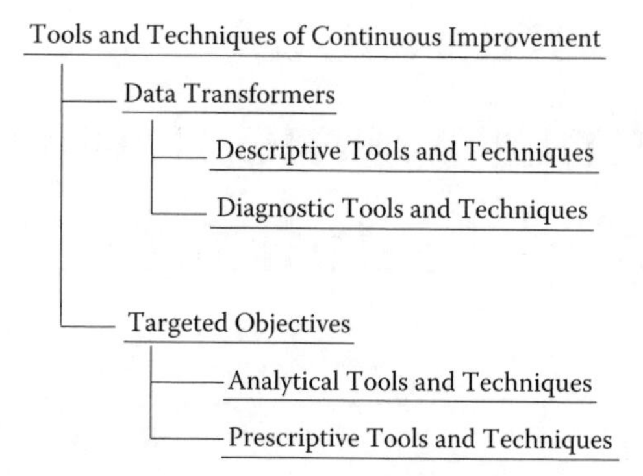

FIGURE 13.1 Data transformers and targeted objectives.

13.3 TARGETED OBJECTIVES

13.3.1 Analytical Tools

Analytical tools are used to dissect inner relationships; for example, when a home builder has a collection of unique and independent data regarding the various options selected by home buyers over a recent, specified time period and the builder wants to identify in which natural groupings each option belongs, it is important to select the most appropriate tool to get the job done. An example of two important, related natural groupings would be that home buyers who select upgraded options in their kitchens tend to do the same in their laundry rooms and bathrooms.

13.3.2 Prescriptive Tools

Prescriptive tools are used to provide direction; for example, when a home builder has a number of choices and needs to select the most preferred choice, the builder might decide to use a tool such as paired comparisons or prioritization matrix. This could occur when a home builder is selecting its primary framer from a large group of frames.

13.4 TOOLS AND TECHNIQUES

For the sake of convenience of both learning and application, the most commonly used tools and techniques of continuous improvement (CI) have been grouped into four major categories: (1) seven quality control (7-QC) tools; (2) the seven management and planning (7-MP) tools, (3) the seven supplemental (7-SUPP) tools, and (4) the seven team (7-TEAM) tools. Sections 13.4.1 through 13.4.4 introduce these categories and the seven tools contained in each category. The remainder of this chapter provides a succinct discussion of these 28 tools as well as some additional but less frequently employed ones, all presented in alphabetical order.

13.4.1 Seven Quality Control Tools

The 7-QC Tools are listed and described in Figure 13.2.

Which of the 7-QC tools should be applied when your team wants to achieve each of the following?	The tool of choice should be the following:
Graphically collect data, either attribute or variable, relative to its frequency of occurrence.	Data tables (check sheets or tally sheets)
Graphically record brainstorming the potential cause(s) of a selected effect, either positive or negative.	Cause and effect analysis
Graphically plot frequency of occurrence of continuous (variable) data within specific class intervals.	Histograms
Graphically plot frequency of occurrence of discrete (attribute) data from the greatest value on the left to the least value on the right.	Pareto analysis
Graphically plot bivariate data points (x, y) to determine if one variable is mathematically related to the other.	Scatter analysis
Graphically plot data, either attribute or variable, as averages or individuals, to ascertain the presence of positive or negative trends.	Trend analysis (graphs or run charts)
Graphically plot data, either attribute or variable, as averages or individuals, to ascertain if a process is in statistical control.	Control charts

FIGURE 13.2 Seven quality control tools (7-QC).

13.4.2 Seven Management and Planning Tools

The 7-MP tools are listed and described in Figure 13.3.

13.4.3 Seven Supplemental Tools

The 7-SUPP tools are listed and described in Figure 13.4.

13.4.4 Seven Team Support Tools

The 7-TEAM tools are listed and described in Figure 13.5.

13.5 ALPHABETIZED LIST OF CONTINUOUS IMPROVEMENT TOOLS AND TECHNIQUES

13.5.1 Affinity Analysis

Affinity analysis (i.e., the use of affinity diagrams) is one of the 7-MP tools. This popular tool is used when a need exists to identify major themes or concepts from a large number of ideas, opinions, or issues. It provides an easy-to-understand, easy-to-use approach to group items that are naturally related and then to identify the one theme or concept generic enough that it ties each grouping together.

The planning process begins with the collection of a large set of qualitative data regarding ideas, opinions, perceptions, desires, and issues. Initially the relationships among these data will not be clear, although the team may detect a sense of how they are related.

Which of the 7-MP tools should be applied when your team wants to achieve each of the following?	The tool of choice should be the following:
Reduce a collection of ideas to a much smaller quantity of themes.	Affinity analysis
From a collection of themes, determine the input and output relationships, and then identify which themes are drivers and which are outcomes.	Interrelationship digraph (ID)
Compare one collection of ideas with another to determine the extent of their interrelationships.	Matrix analysis
Compare the attributes of a collection of alternative solutions using a selected set of criteria to determine which of the solutions is preferred to achieve a particular objective.	Prioritization matrix
Separate a top-level idea into its next-level component parts and then divide these into their next-level component parts, etc.	Tree diagram
Given an idea, identify the various possible approaches that might achieve the idea, with each approach to include requirements, potential failures, possible countermeasures, and the likelihood of success or failure for each countermeasure, thus resulting in determination of the optimal approach to achieve the idea.	Process decision program chart (PDPC)
Sequentially plot a group of events using series and parallel relationships to portray the network of events with its critical path as well as the start and finish times for each event.	Activity network diagram (AND)

FIGURE 13.3 Seven management and planning tools (7-MP).

Affinity analysis is a creative process, as opposed to being a logical process; it helps to generate consensus by sorting written documents in lieu of a discussion of ideas. These documents can be note cards or self-sticking note papers. The latter are preferred for use on vertical surfaces such as walls, whiteboards, or windows.

Affinity analysis is the appropriate tool to use in the following circumstances:

- When chaos exists
- When a team is overwhelmed by a surplus of ideas
- When breakthrough thinking is needed
- When broad issues or themes must be identified

Figure 13.6 is demonstrates the application of affinity analysis.

13.5.2 Activity Network Diagram (AND)

Developed in Japan in the 1980s, the AND is a derivative of the critical path method (CPM) and the program evaluation and review technique (PERT) originated in the United States in the 1950s. CPM resulted from a joint project by DuPont and Remington Rand Univac in 1957. The goal of the project was to determine how to

Which of the 7-SUPP tools should be applied when your team wants to achieve each of the following?	The tool of choice should be the following:
Use an inspection or test form to tally quantities of various types of defects as well as to specify their locations relative to unit geometry.	Defect map
Create a detailed chronological listing of those process-related events which differ from the norm and which is to be used as a problem-solving tool when a control chart, histogram, or Pareto analysis indicate the presence of an abnormality.	Events log
Use an analytical concept to categorize data drawn from a heterogeneous source into more homogeneous classes.	Data stratification
Apply a formalized administrative process to ensure that each and every unit in a population has an equal opportunity for selection or inclusion in a sample. The team can use random number tables, computer-generated random numbers, and/or 10/20-sided dies.	Randomization
Graphically depict a process, linear or nonlinear, indicating the sequence of steps as well as the input and output for each step. The flow chart portrays only the process steps while the map indicates location, area, or department as well.	Process flow chart/map
Incorporate the use of a display board located in proximity to a specific process. This display board contains a current control chart, process flow chart, Pareto analysis (for attribute data) or histogram (for variable data), events log, and any other process-related data.	Progress center
Use a statistically based procedure to determine sample size and to select units from a population of units for purposes such as inspection and test.	Statistical sampling

FIGURE 13.4 Seven supplemental tools (7-SUPP).

best reduce the time to perform routine plant overhaul, maintenance, and construction work. PERT was first used by the U.S. Navy in the design and production of the Polaris ICBM.

The AND is a tool for determining the optimal time for accomplishing a task and for graphically displaying the flow of activities that lead to its achievement. An AND is most effective when the activities for a task are well known and there is a high degree of predictability and confidence in that knowledge. Each activity is plotted in sequence, the time to accomplish each activity is annotated on the plot, and then the times are used to determine the earliest and latest possible times that any given activity can begin.

An AND clearly indicates which activities can be performed in parallel (concurrently) and which must be accomplished in series (sequentially). In this way, the optimal activity flow can be determined, and the minimum and maximum times to complete the entire task can be calculated. If the activities are not well known and

Which of the 7-TEAM tools should be applied when the team wants to achieve each of the following?	The tool of choice should be the following:
Compare items on a list using a standard or benchmark previously agreed to by the team.	Forced choice
Prioritize a brief list of ideas by evaluating each idea on the list with every other idea.	Pairwise ranking
Determine the highest priority ideas within a long list of items, but the team is large and needs to avoid any "win–lose" situations.	Multivoting
Reduce a long list of ideas to a shorter, more manageable size or, alternatively, assign operational priorities to each item on a long list.	List reduction
Avoid any of the following: • Pressure toward team conformity • Effect of senior status • Going off on tangents • Getting too narrowly focused • Being unwilling to say something that sounds completely foolish	Nominal group technique (NGT)
Organize a large number of ideas using the creative rather than the logical part of the brain by working with colors and images to promote visualization of ideas.	Mind mapping
Reach consensus on subjects of consequence by merging ideas, conclusions, or beliefs whether participants are in the same room or geographically dispersed and which enables team members to participate anonymously.	Delphi method

FIGURE 13.5 Seven team tools (7-TEAM).

understood, a high level of frustration can result. When this is the case, the process decision program chart (PDPC) is a better tool for graphically displaying the flow of activities leading to a task or a project. Figure 13.7 provides an example of an AND.

13.5.3 Bar Chart

A bar chart is a form of graphical presentation designed to quickly and simply communicate quantitative information. It always has two dimensions: (1) the horizontal or x-axis, which is traditionally used for an independent variable; and (2) the vertical or y-axis, which is used to describe the dependent variable.

The independent variable can be either variable/continuous data or attribute/discrete data. A histogram is a unique version of a bar chart that presents variable/continuous data as the independent variable. A Pareto diagram is also a bar chart, but one that uses attribute/discrete data as the independent variable. Both histograms and Pareto diagrams use the vertical axis to communicate about the dependent variable, which is usually frequency of occurrence or cost.

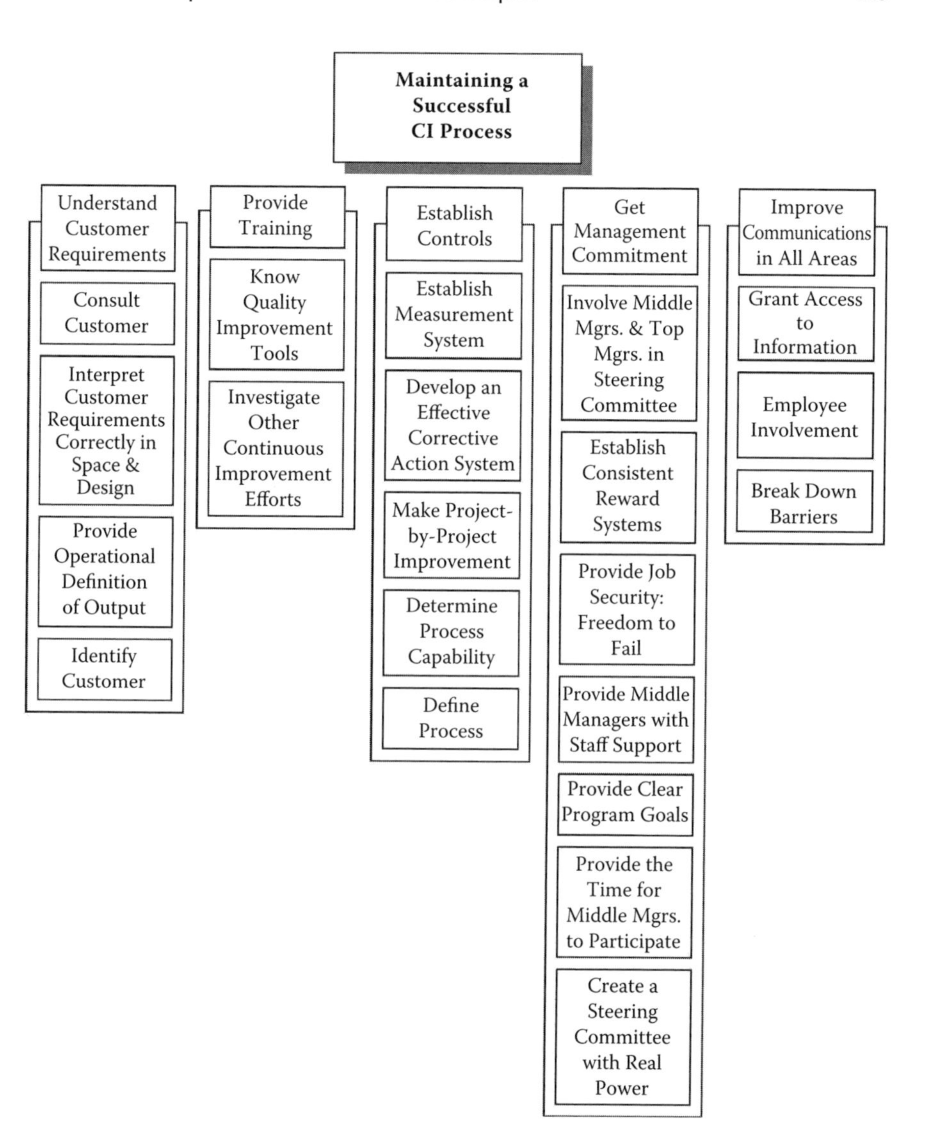

FIGURE 13.6 Affinity analysis.

A bar chart can be displayed with the bars oriented either vertically or horizontally, depending on the inclination and needs of the chart master.

13.5.4 Box and Whisker Chart

A box and whisker chart (also known as a boxplot) is used to simultaneously display the median (one of three measures of central tendency; the other two are mean/average and mode) and the quartile (measure of dispersion). This presents a more complete picture of the status of a process than the traditional line/run chart.

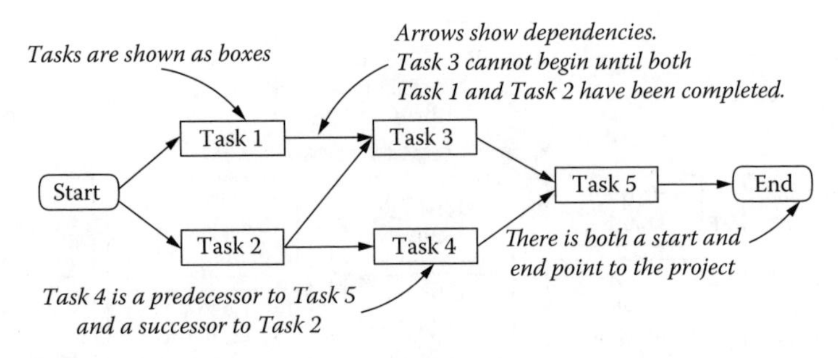

FIGURE 13.7 Activity network diagram (AND).

The following is an example of what can be done with a data set that may not be normally distributed other than using the mean and standard deviation to create a frequency distribution:

Data Pool (n = 20)

Quartile			
First	**Second**	**Third**	**Fourth**
50	44	**38**	34
49	43	38	33
47	42	37	32
45	41	36	31
45	**40**	35	30

Md = 39

In a box and whisker chart, the top of the box is the data point at the top of the second quartile, and the bottom of the box is the data point at the bottom of the third quartile. In the case of the example data pool, the top of the box would correspond to a value of 44, the center of the box would correspond to the median (39), and the bottom of the box would have a value of 35. The far end of the top whisker would be 50, and the far end of the bottom whisker would be 30, as in Figure 13.8.

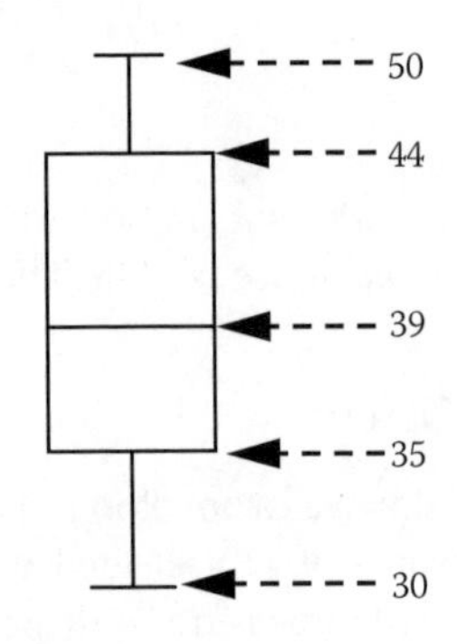

FIGURE 13.8 Box and whisker chart.

13.5.5 Brainstorming (Classical versus 635 versus Imaginary): Classical Brainstorming

The basis of subjectively based decisions is the generation of creative alternatives. The purpose of classical brainstorming is to place as many ideas as possible on the table where they can be discussed and evaluated and by drawing on the extensive knowledge and experience of team members.

Classical brainstorming is a technique used to generate multiple ideas by a team. Similar to the nominal group technique (NGT), a methodology covered later in this chapter, it is useful to identify problems, to determine their causes, and then to generate solutions to the problems. Properly conducted, a classical brainstorming session will produce a maximum quantity of potential ideas in a minimum amount of time by fully using the team's collective creativity.

Classical brainstorming has been used successfully to identify problems, to establish organizational goals and objectives, to construct process flowcharts and process maps, to expedite cause and effect analysis, and to search out potential corrective actions.

The basic ground rules of classical brainstorming are quite straightforward:

- No criticism
- Get crazy
- Quantity, not quality
- Group effort

The following section describes interrelationships between these ground rules.

13.5.5.1 Interrelationships between Ground Rules of Classical Brainstorming

Brainstorming 635: Brainstorming 635 is based on how the process is conducted. There are usually six people who write three ideas and who pass the ideas around their circle about every five minutes or so. During the process of idea generation, the participants take ideas produced by other team members and attempt to further develop them.

Brainstorming 635 can be used when a team needs another technique for producing additional ideas. It is a viable alternative to classical brainstorming when the personalities or team dynamics require a different approach because team members come from different levels of an organization; it is also effective when a problem requires additional analytical thinking. Brainstorming 635 is especially recommended when the number of participants makes classical brainstorming impractical. With a large number of persons, several 635 groups can be established, or the Brainstorming 635 worksheets can be passed a lesser number of times.

Brainstorming 635 is accomplished without discussion or conversation of any kind. Team members write down three ideas and pass their papers to the person on their right (only for the sake of tradition). They read these silently and add three more ideas that are triggered by the preceding ideas. This continues around their circle until each person gets back his or her original paper. Brainstorming 635 uses a silent, iterative approach and encourages more reflection than does classical brainstorming. Using this version, Brainstorming 635, the potential exists for 108 ideas to be generated in 30 minutes.

Imaginary brainstorming: The unique approach used by imaginary brainstorming encourages the generation of unusual ideas. To begin, classical brainstorming is conducted, and then a single item in the original brainstorming description is altered. This is followed by the generation of new ideas based on the altered description. The results of the imaginary brainstorming are then compared with the original results. This is portrayed in Figure 13.9.

Imaginary brainstorming helps break established patterns of thinking, thus allowing a team to identify more creative ideas and solutions to problems.

13.5.6 Cause and Effect Analysis

The original concept of brainstorming as a means to generate and collect possible causes of a specific effect (either positive or negative) offers minimal structure to

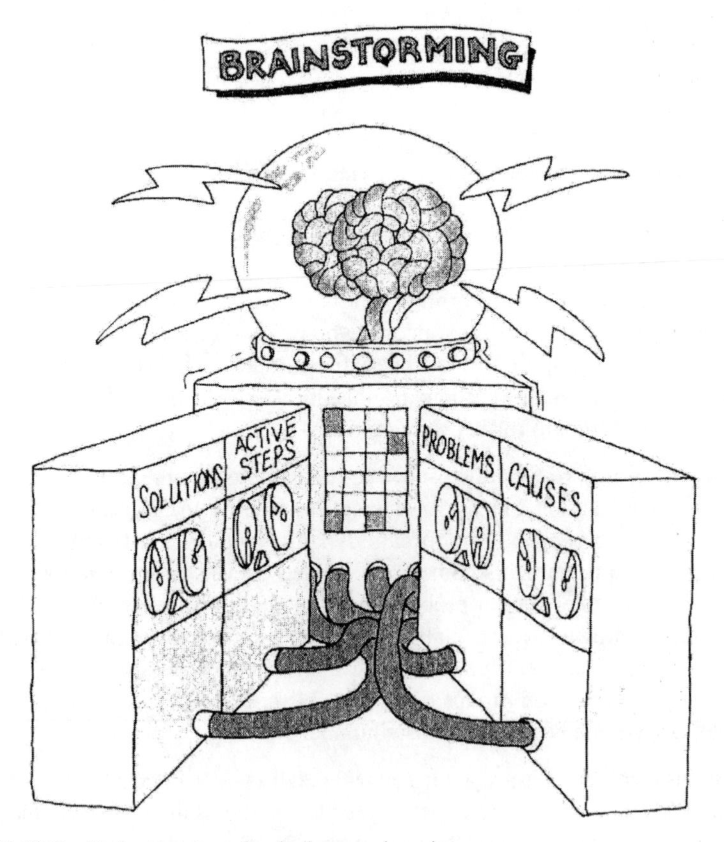

FIGURE 13.9 Brainstorming: classical versus imaginary.

assist members of a cross-functional team. Cause and effect analysis was developed in the 1950s by Dr. Kaoru Ishikawa to provide the structure missing from traditional brainstorming.

Cause and effect analysis is sometimes referred to as the Ishikawa diagram but perhaps more frequently as the fishbone diagram because of its similarity in appearance to the skeleton of a fish. As a team generates possible causes or ideas using brainstorming, these ideas are placed on the diagram by a team leader or facilitator. This contributes to more efficient analysis and evaluation of the causes by graphically establishing the relationships between and among these various ideas. The effect is generally described as a problem to facilitate the brainstorming of the potential causes.

Cause and effect analysis helps team members to identify which factors could, either directly or indirectly, contribute to the effect being studied. Usually one or more of the "5 Ms and an E" are the primary sources or causes of the effect (Figure 13.10):

- Men/women
- Machine
- Measurement
- Material
- Method
- Environment

13.5.7 Check Sheet and Checklists

Check sheets, or checklists as they are sometimes known, provide a systematic method for collecting and displaying specific data. In most cases, check sheets or checklists are simply forms designed for the purpose of collecting specific data. They provide a consistent, effective, and economical approach to gathering data, organizing it for analysis, and displaying it for preliminary review.

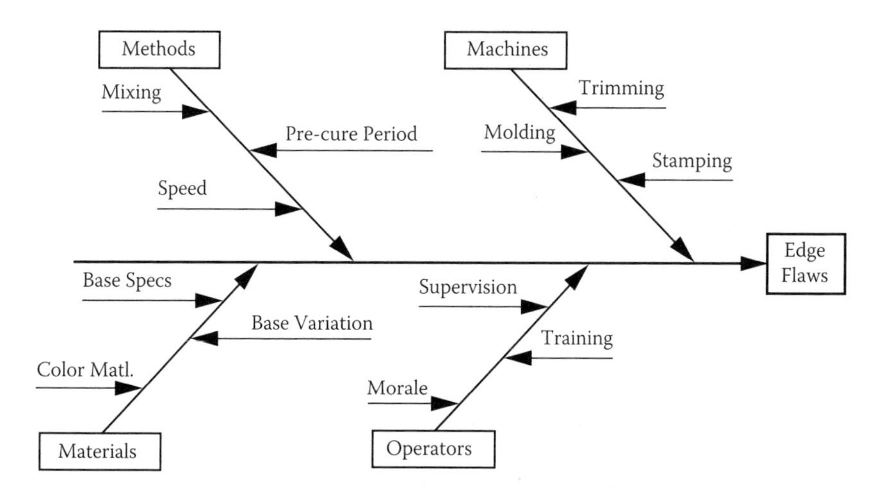

FIGURE 13.10 Cause and effect analysis (fishbone diagram).

Check sheets or checklists sometimes take the form of manual check sheets where automated data are not necessary or available. Data figures and check sheets should be designed to minimize the need for complicated entries that encourage personnel error (the reader should review error proofing later in this chapter). Properly designed, check sheets or lists can facilitate development of data summaries for other tools such as Pareto analysis and histograms. Simple, straightforward tables are a key to successful data gathering.

13.5.8 Control Charts

When processes require close surveillance of specific attributes as well as critical dimensions or rates during production, this status assessment is best accomplished with the application of statistically based control charts. The primary purpose of control charts is to graphically show trends in the frequency of occurrence of the attributes or in the magnitude of the dimensions or rates with respect to maximum and minimum limit lines (referred to as upper and lower control limits).

A process that has reached a state of statistical control has predictable variation and output. Control charts provide a means of anticipating and correcting whatever special causes, as opposed to common causes, may be responsible for the generation of defects or unacceptable variation.

Control charts are used to direct the efforts of engineers (e.g., design, process, quality, manufacturing, industrial), operators, supervisors, and other process improvement team (PIT) members toward special causes when, and only when, a control chart detects the presence and influence of a special cause. The ultimate power of control charts lies in their ability to separate out special (assignable) causes of defects and variation.

The objective of control chart analysis is to obtain evidence that the inherent process variability (known as the range, or R) and the process average (known as the average of the averages, or X-double bar) are no longer operating at stable (predictable) levels. When this occurs, one or both are out of control (not stable); thus, appropriate and timely corrective action should be introduced. The purpose of establishing control charts is to distinguish between the inherent random variability of a process (known as common causes) and the variability attributed to an assignable (or special) cause. Common causes are those that cannot be readily changed without significant process reengineering.

Control charts are applied to study specific ongoing processes to keep them operating with minimum defects and variation. This operating state is known as Six Sigma and can be verified using performance measures such as C_p, the process capability index, and C_{pk}, the mean-sensitive process capability index. This is in contrast to downstream inspection and test, which aim to detect defects and variation after they have been generated by the process. In other words, control charts are focused on prevention rather than detection and rejection.

It has been confirmed in practice that cost and efficiency are better served by the prevention concept than by the detection concept. It costs as much to make a bad part as a good part, and the cost of applying control charts is repaid many times over as a direct result of improving production quality.

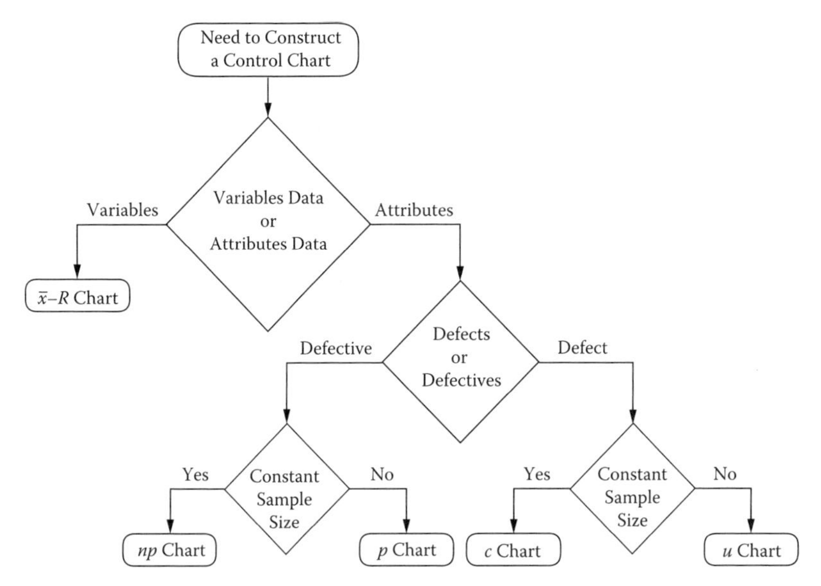

FIGURE 13.11 Control chart applications. Defect = A failure to meet one part of an acceptance criteria. Defective = A unit that fails to meet acceptance criteria due to one or more defects.

13.5.8.1 Types of Control Charts

Just as there are two types of data (i.e., continuous and discrete), there are also two types of control charts: (1) variables charts for use with continuous data, and (2) attributes charts for use with discrete data.

Variables control charts should be used whenever measurements from a product or process are available. The following are examples of continuous data: lengths, diameters, time duration values, temperatures, and electrical performance values such as voltages and amperages.

Whenever possible, variables control charts are preferred to attribute control charts because they provide greater insights regarding product or process status. When continuous data are not available but discrete data are, then it is appropriate to use attributes control charts. There are only two levels for attribute data (i.e., conform/nonconform, pass/fail, go/no-go or present/absent; however, they can be counted, recorded, and analyzed. Examples of discrete data include presence or position of a label; installation of all required fasteners; presence of excess or insufficient black paper for leak-proofing roofs, windows, and doors; and measurements when recorded as accept/reject.

Figure 13.11 indicates when various control charts should be applied.

13.5.9 Data Stratification

The purpose of data stratification is to convert a heterogeneous population (mixed types of units) into a collection of homogeneous subpopulations (all the same or similar types of units). This separation process facilitates whatever studies or analyses of the heterogeneous population from which statistical samples may be drawn. Examples of data stratification include analyzing a population of homes to determine which types have specific kinds of defects or excessive variation, examining a population of defects created by a process to ascertain the "critical/vital few" categories, and studying a population of employees to identify the needs and expectations of each category of employees.

The procedure for data stratification is not complicated. Once the population of concern has been identified, it is examined to determine the various types of categories that exist therein, such as size (dimensions), age, vendor/supplier (sources), color, weight, distance, gender, and cost. Next, the population is stratified (divided) according to the pertinent categories. Then, finally, as data regarding the population are collected, the categorical information about the sampled units is recorded using a tally sheet or some other type of data table. Home builders frequently stratify data about the homes they build by square footage, number of stories, and classification (entry level, move up, and executive level).

13.5.10 Defect Map

When patterns of specific defects or excessive variation occur over time, it is desirable to collect and log these patterns as a series of visual records. These records or defect maps are used to record the locations of the defects relative to the unit geometry. Defect maps provide much-needed assistance in problem solving; that is, the focus moves to more detailed, lower-level problems that are easier to identify and correct. They are readily applied to both fabrication and assembly defects as well as to supplier problems.

The procedure is straightforward: Create a clearly defined product sketch with identifying coordinates, collect and reduce the resulting data to usable statistics, and then focus subsequent problem solving on the critical few (i.e., the worst) first. Figure 13.12 offers a template for a home defect map.

13.5.11 Events Log

The sole purpose of an events log is to relieve process superintendents and other support personnel from the necessity of recalling process changes when defects or excessive variability is detected. The detection may come about as a result of using control charts, Pareto analysis, or histograms.

The original events logs were created in response to demonstrated needs for process status documentation. Entries on an events log should be made by any process-related personnel whenever process changes involving the 5 Ms and an E (i.e., men/women, methods, measurement, machine, material, and environment—from cause and effect analysis) occur. Events log entries are analyzed whenever control chart, Pareto diagram, or histogram data indicate significant alterations to process output, either good or bad.

13.5.12 Error or Mistake Proofing (*Poka-Yoke*)

Originally known in Japan as *Baka-yoke*, which means something akin to providing protection from crazy or foolish persons, error proofing or *poka-yoke* (pronounced

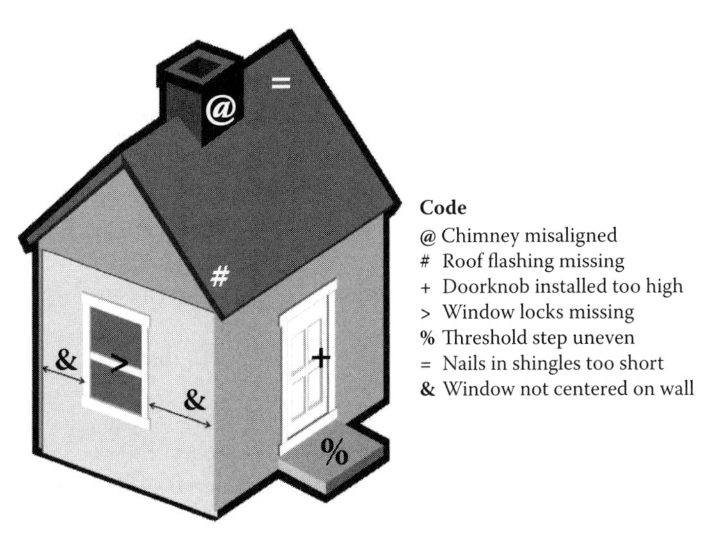

FIGURE 13.12 Defect map.

POH-kah YOH-kay) is a means to provide "fail-safe" protection from human errors. There are many types of error proofing in the form of simple machine or process revisions and modifications as well as product variations and retrofits.

The following sections provide examples of error proofing at various levels.

13.5.12.1 Basic

- Use cross-checking when totaling a number of columns.
- Write down special or general instructions to employees for their future reference.
- Increase the effectiveness of internal communications. Ask employees to repeat instructions to ensure understanding.

13.5.12.2 Intermediate

- Use different colored paper for different purposes. This helps to ensure correspondence is directed to the correct location.
- Make certain that on–off switches on your organization's PCs are physically distanced from other switches or buttons on the PCs so they are not easily switched off with a resulting loss of data or information.
- Put all letters in envelopes with plastic windows to display names and addresses already typed on the letters. This serves the dual purposes of elimination of retyping names and addresses as well as letters being sent to the wrong persons.

13.5.12.3 Advanced

- When using polarized electrical equipment, use electrical plugs that cannot be inserted into electrical sockets without the necessary orientation. This is accomplished by making one of the plug inserts wider or narrower than the other.

- When using machinery that has a potential to cause harm to its operator, incorporate safety features into the operational design of the machine such that necessary guards must be in place before the machine can function.
- When starting an automobile or truck, it could accidentally move in an unexpected direction with disastrous results. To eliminate the potential for such events, incorporate a safety feature that prevents starting a vehicle and putting it in gear without first depressing the brake pedal.

13.5.13 Five Whys

The five whys is a technique for discovering the root cause (or causes) of a problem by repeatedly asking the question, "Why?" Five is an arbitrary figure; you may need to ask the question a few times, more or less, but for certain it will be more than once. You never know in advance exactly how many times you'll have to ask why. The five whys technique helps to identify the root causes of a problem and to see how different causes of a problem might be related.

To begin, the problem should be described in very specific terms. Then, ask why it happens. If the answer doesn't identify a root cause, then ask why again. The root cause has been identified when asking why doesn't provide any more useful information. Continue asking why until the root causes have been identified. As previously noted, this can take more or less than five whys.

It's important to remember to always focus on the process aspects of a problem rather than on the persons involved. Finding scapegoats doesn't solve problems! Here's an example. An operations department wanted to find out why it missed its initial operating capability (IOC) date:

Statement: We missed our IOC!

Why? (#1)
Response: Our contract delivery schedule slipped.
Why? (#2)
Response: There were a lot of engineering changes.
Why? (#3)
Response: The contractor didn't understand our initial requirements.
Why? (#4)
Response: Our technical data package wasn't prepared very well.
Why? (#5)
Response: We took only one week to prepare it.

At this point, the group members recognized poor requirements planning as a root cause of their problem. As a result, they decided to allow more time up front in the planning process for requirements analysis.

Suppose you hear the following statement: "We used to have a respectable defect rate; it ran about 100 parts per million (ppm)/defects per million defect opportunities (dpmo). Then, a few months ago, it jumped up to about 2,000 ppm."

Using the five whys technique, the subsequent discussion might go something like this:

Q: Do you have any ideas why the big increase in the defect rate?
A: I'm not sure, but it could be a couple of things. For example, it might be the new materials that our new supplier is sending us, or it could be a lack of training with regard to the most recently hired superintendents.
Q: What do you think it is? (Why #1)
A: I think it might be a combination of the new materials and the lack of training.
Q: Why do you think that's the reason for the higher defect rates? (#2)
A: Well, we had a pretty high turnover rate a few years ago, and from time to time the defect rate would jump up and then come back. We were never absolutely sure why.
Q: How about the new materials? Why do you think they might be contributing to the increase in the defect rate? (#3)
A: It's nothing official yet, but we've heard that the new supplier has been cutting corners and that the quality of the new materials we've been receiving from them is questionable.
Q: That all sounds pretty interesting, but why do you think that it might be a combination of the material quality and the higher conveyor speeds? (#4)
A: It's the timing. The defect rate didn't really jump up when the new supers were hired, but it did when the new materials started being used.
Q: What can we do to find out for sure? (#5)
A: We can use a design of experiments (DOE) to figure out if the interaction between the new materials and the training/experience of the superintendents are really why the defect rates have increased.

Ask why 5 times to determine the root cause:

1. Q: Why is this air compressor not running?
   A: Because its drive belt is broken.
2. Q: Why is the drive belt broken?
   A: Because the drive gear was not turning fast enough.
3. Q: Why was the drive gear not turning fast enough?
   A: Because the drive shaft's lubrication reservoir ran empty.
4. Q: Why did the drive shaft's lubrication reservoir run empty?
   A: Because the preventive maintenance (PM) for this machine is overdue by almost two weeks.
5. Q: Why is the PM for this machine overdue by almost two weeks?
   A: Because the lubrication maintenance person is on a two-week vacation.
6. Q: Why didn't someone else cover for the lubrication maintenance person during his vacation?
   A: Because we do not have a vacation coverage plan for the maintenance department, and our operators are not trained and empowered to do lubrication.

Remember: sometimes you have to ask why more than five times to get to the root cause.

13.5.14 Forced Choice

Forced choice is used by teams to compare the items in a list against a standard previously agreed to by team members. Typically, this standard is a value judgment selected by group consensus. Remember, a consensus is a general agreement with which all the team members can live even if it's not exactly what each member would prefer. Some possible standards include cost, quality, time, and the like. The list of items can be generated by brainstorming, which was discussed in its three forms earlier in this chapter. Using forced choice results in the identification of contradictions in logic as well as in prioritization of the list using pairwise ranking, which is discussed later in this chapter. Using a standard question such as, "Which is more expensive, Item 1 or Item 2?" helps to maintain consistency in the comparison process.

13.5.15 Histogram

A histogram is a graphical representation of the distribution of variable data (data generated by taking measurements) or attribute data (data generated by counting) using a bar chart. It is commonly used to visually communicate information about a process or a product as well as to help make decisions regarding prioritization of improvement initiatives.

This information is represented by a series of equal-width columns of varying heights. The columns are of equal width because they all represent a specific class interval within a range of observations. Column height is a function of (i.e., directly proportional to) the number of observations (frequency of occurrence) within the interval covered by each column. Thus, column height varies according to the number of "items" within a specified class interval.

With most naturally occurring data, there is a tendency for many of the observations to occur in proximity to the center of the distribution (known in statistical circles as the *central tendency*) with progressively fewer points occurring farther from the center.

Histograms offer a quick look at data at a single interval in time (e.g., for the last hour, last shift, last day). They do not display variation or trends over time. A histogram displays how the cumulative data look now. It is useful in understanding the relative frequencies (percentages) or frequency (quantity) of the data and how the data are distributed.

Many candidate processes or products for improvement can be identified using this basic tool. The frequency and shape of the data distribution provide insights that would not be apparent from data tables (check sheets or lists) alone. Histograms also form the basis for two other frequently used total quality management (TQM)/continuous improvement (CI) tools: Pareto analysis and C_p. Figure 13.13a presents a typical table of continuous data, while Figure 13.13b displays an associated histogram.

Data Range	Frequency
0–9	1
10–19	3
20–29	6
30–39	4
40–49	2

(a)

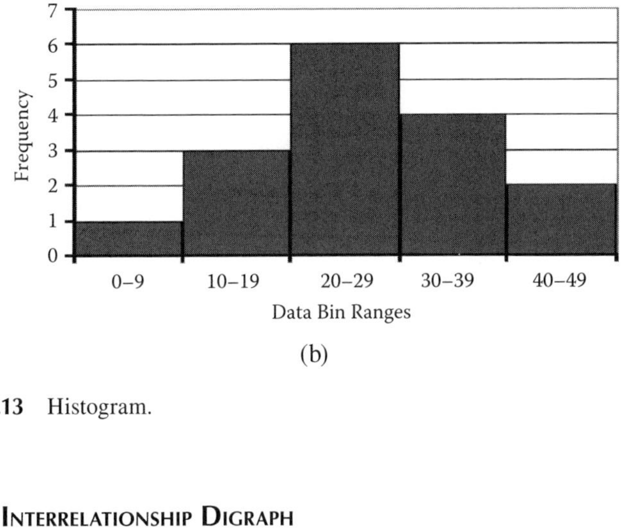

Frequency
7
6
5
4
3
2
1
0
0–9 10–19 20–29 30–39 40–49
Data Bin Ranges

(b)

FIGURE 13.13 Histogram.

13.5.16 Interrelationship Digraph

The relationships among qualitative data elements are not linear; in fact, they are often multidirectional. In other words, an idea, a concept, an issue, or an action can affect more than one other idea, concept, issue, or action, and the magnitudes of these effects can and do vary. Additionally, the relationships are often hidden or not clearly understood.

The interrelationship digraph (ID) is an effective tool for clarifying the relationships among ideas, concepts, issues, or actions. This is accomplished by graphically mapping the cause and effect linkages (i.e., the sequential connections among them). The usual input for an ID is the output or result generated from using other tools such as affinity analysis, brainstorming, cause and effect analysis, or tree diagrams. The ID can, however, be used to analyze a set of ideas, concepts, issues, or actions generated without first using another tool.

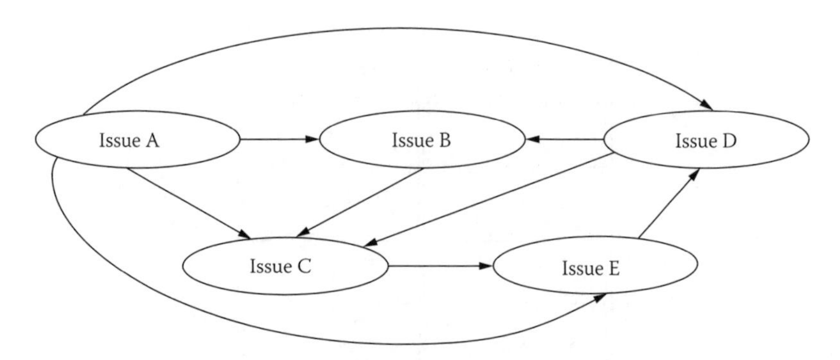

FIGURE 13.14 Interrelationship diagraph-original (arrow) method).

IDs are used in the following cases:

- When root causes must be identified
- When a number of interrelated items requires better definition
- When data cannot be easily, quickly, or inexpensively acquired to identify root causes
- When scarce resources demand a carefully focused effort

There are two approaches to using an interrelationship digraph. The original approach, referred to as the arrow method, was developed in Japan and uses a series of arrows to graphically portray known or existing interrelationships. This method works quite nicely for a relatively small group of data elements. As the number of elements begins to grow, however, the arrowheads tend to overlap each other, causing a confusing picture not unlike a plate of spaghetti. Figure 13.14 introduces the original and most used approach to the ID.

The organization GOAL/QPC teaches an alternate approach referred to as the matrix method. This approach was specially created to deal with situations that involve a medium to large number of data elements. The matrix method employs an L-shaped matrix that methodically organizes the multitude of interrelationships among the elements. Figure 13.15 presents the GOAL/QPC approach to using the ID tool when 10 or more data elements require analysis.

Both approaches are designed to facilitate the subsequent analysis to prioritize the elements, that is, to determine the critical few that have the greatest impact on achieving whatever results are desired.

13.5.17 Line Graph

Line graphs compare two variables. Each variable is plotted along a straight line axis. A line graph has a vertical axis and a horizontal axis. So, for example, if you wanted to graph the time required to build a specific model home, you could put home number (e.g., 1, 2) along the horizontal or x-axis and number of workdays to complete each home along the vertical or y-axis.

Issues	A	B	C	D	E	In	Out	Priority
A		↑	↑	↑	↑	0	4	1
B	←		↑	←		2	1	3
C	←	←		←	↑	3	1	3
D	←	↑	↑		←	2	2	2
E	←		←	↑		2	1	3

FIGURE 13.15 Interrelationship diagraph-matrix method.

Some of the strengths of line graphs are as follows:

- They are good at showing specific values of data, meaning that given one variable the other can easily be determined.
- They show trends in data clearly, meaning that they visibly show how one variable is affected by the other as it increases or decreases.
- They enable the viewer to make predictions about the results of data not yet recorded.

Unfortunately, it's possible to manipulate the way a line graph appears to make data look a certain way. This is accomplished either by not using consistent scales on the axes, meaning that the value in between each point along the axis may not be the same, or when comparing two graphs using different scales for each. It is important to be aware of how graphs can be made to look a certain way, when in fact that might not be the way the data really are. Figure 13.16 offers some examples of line graphs.

13.5.18 List Reduction

List reduction is a technique used by team or group facilitators to reduce a large number of ideas to a much smaller, more manageable list, usually three to five items. It is a structured series or rounds of voting designed to assist cross-functional teams of subject matter experts (SMEs) to deal with an unmanageable list of topics. Alternatively, list reduction can be effectively used to assign operational priorities to each item in a list of many items.

Whether the team is small (three to five persons) or large (over 10 persons), this technique is ideally suited to bring closure to discussions regarding which topics are most and least important. As in any situation when a facilitator is leading the discussion, it is strongly recommended that a disinterested, unbiased third party be used to keep the team or group on track.

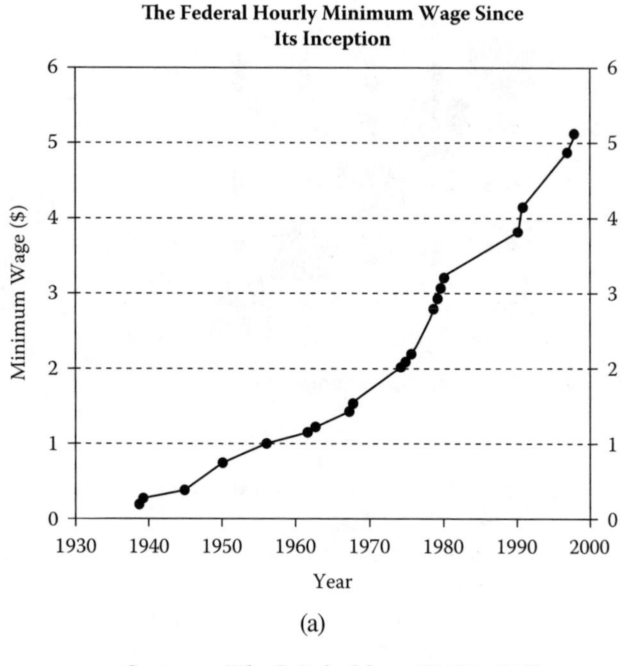

(a)

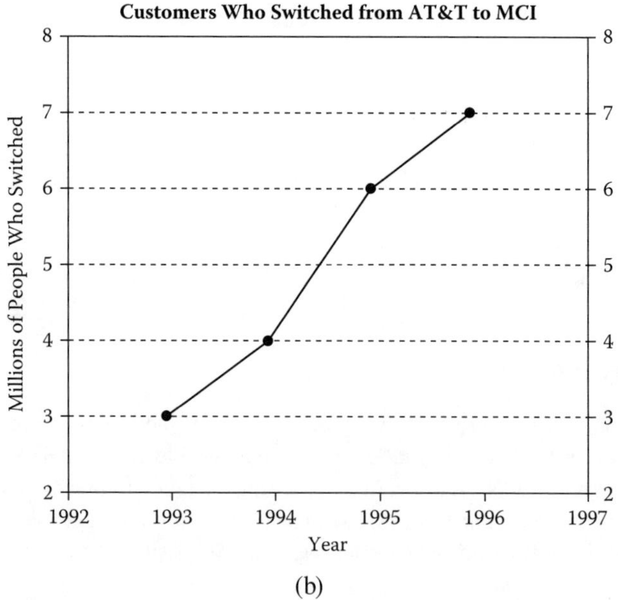

(b)

FIGURE 13.16 (a) Line graph: example A; (b) line graph: example B; (c) line graph: example C.

13.5.19 Matrix Analysis

Matrices or matrix diagrams are used by cross-functional teams of SMEs to organize and compare two or more sets of data elements (i.e., ideas, concepts, issues, or actions).

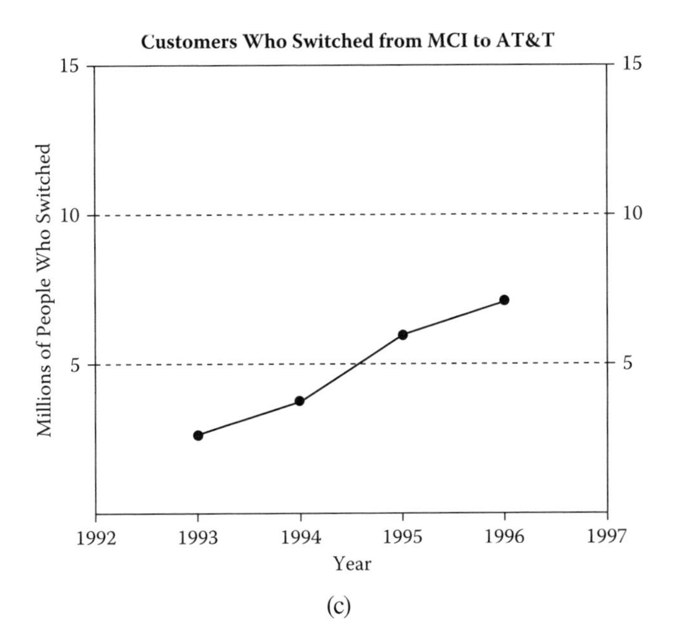

FIGURE 13.16 *(Continued).*

The procedure is simple enough. First, arrange two sets of data on an L-shaped framework or matrix with one set, usually the independent variable, on the vertical axis and the other, usually the dependent variable, on the horizontal axis. Then examine the connecting points or cells for each row–column combination to determine the extent of whatever relationship exists. The resulting matrix diagram reveals all the relationships among the elements and visually clarifies the influence each element has on every other element.

Matrices come in a variety of dimensions and shapes. Two-dimensional matrices include the L, T, Y, and X shapes. A three-dimensional matrix has a C shape. The L matrix is used to compare two sets of variables, whereas the T, Y, and C matrices are used for three sets of variables, and the X matrix for four.

Matrix diagrams are employed in the following instances:

- When definable and assignable tasks must be deployed within an organization
- When selected activities need to be compared with other activities
- When an organization wants to prioritize its current operations versus potentially new operations

Figure 13.17 portrays the various matrices just noted.

13.5.20 Mind Mapping

The key to breaking old paradigms is to use the intuitive powers of the mind. Mind mapping is a powerful technique for invoking the creative part of the brain. It works with colors and images to promote the visualization of ideas rather than evaluation via logic. It is a tool for organizing a large number of ideas and, as such, should be considered for use with affinity analysis, one of the 7-MP tools. In addition, mind mapping has been successfully used before the start of a brainstorming session.

It is usually performed by a small group working in silence around a common map. In larger groups, it is recommended that one person be the recorder to draw the map while the others provide the images. Mind mapping begins when the participating group or team clearly defines a specific topic. A symbol of the topic, referred to as the central image, is drawn in the center of a piece of paper. For the sake of clarity, a word or two may be placed next to the symbol. Then, either in silence or with a background of relaxing music, the members of the group concentrate on the topic.

Y

X		A	B	C	D	E
	1					
	2					
	3					
	4					
	5					

(a)

Z	k	j	i	h	g	f	e	d	c	b	a		A	B	C	D	E	F	G	H	I	J	K	Y
												1												
												2												
												3												
												4												
												5												
												6												
												7												
												8												
												X												

Z

e	d	c	b	a		X
					1	
					2	
					3	
					4	
					5	

Y

X		A	B	C	D	E
	1					
	2					
	3					
	4					
	5					

(b)

FIGURE 13.17 (a) L-shaped matrix; (b) T-shaped matrix; (c) Y-shaped matrix; (d) X-shaped matrix; (e) C-shaped matrix.

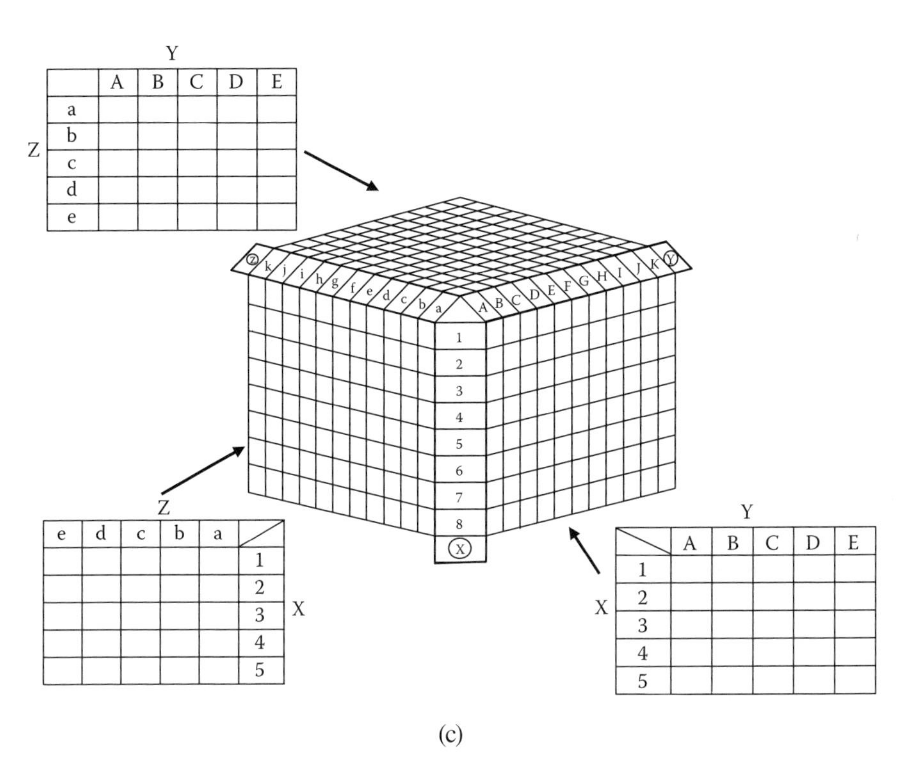

(c)

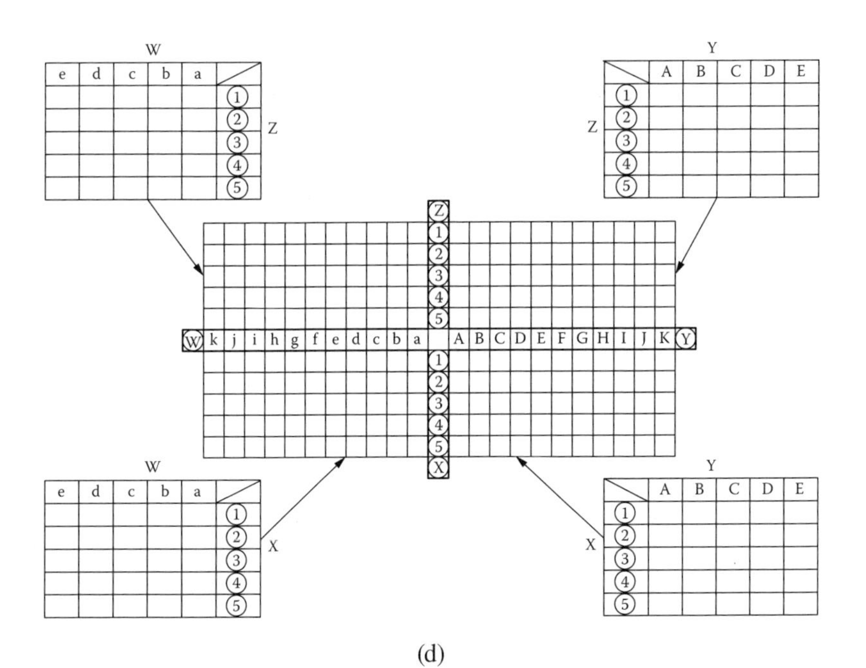

(d)

FIGURE 13.17 *(Continued).*

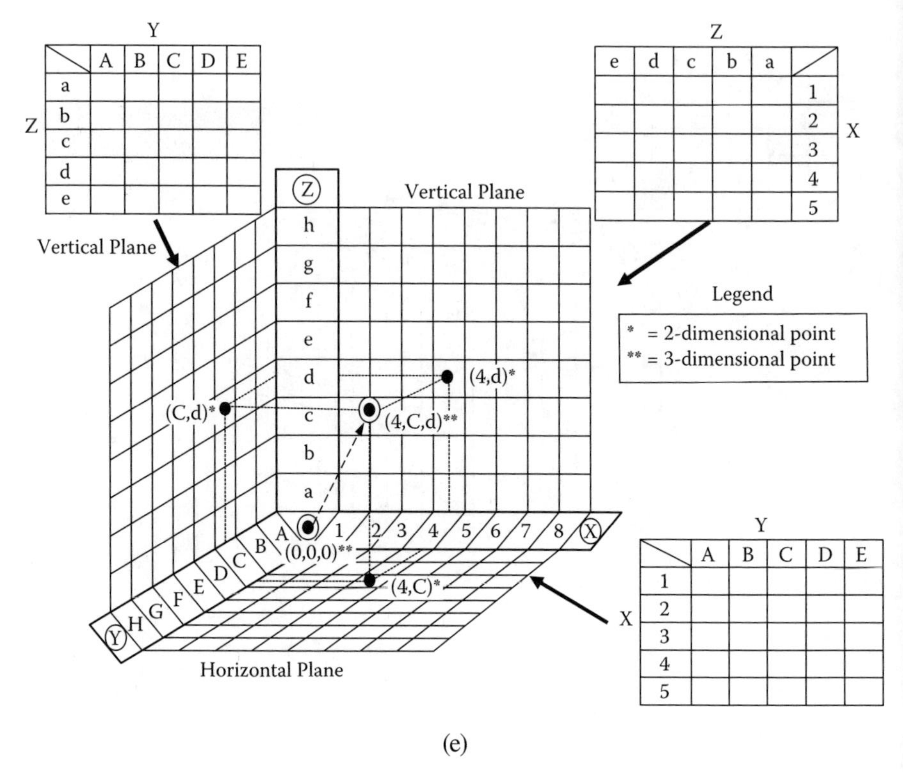

FIGURE 13.17 (*Continued*).

As ideas or images come to mind, they are recorded around the central image. Colors and symbols are used to stimulate the creativity of the team members. One word may be added for clarity. This is followed by using a line to connect each image to the central image. At this point the team members focus on each new image and capture the additional images it stimulates. This process continues until the team runs out of ideas. Finally, ideas with common themes are grouped by drawing a colored line around them, marking them with a common code, or redrawing the map to cluster them together.

Taking frequent advantage of mind mapping will improve a team's ability to involve the creative right brains of the team's members in decision making. Figure 13.18 is an example of a mind map.

13.5.21 Multivoting

Multivoting is a quick and easy way for a group or team to determine the highest priority items within a list of items. It is best suited for large groups and long lists. Its simplicity helps a team to achieve the following:

- To prioritize a large list without creating a win–lose situation in the group or team that generated the list.

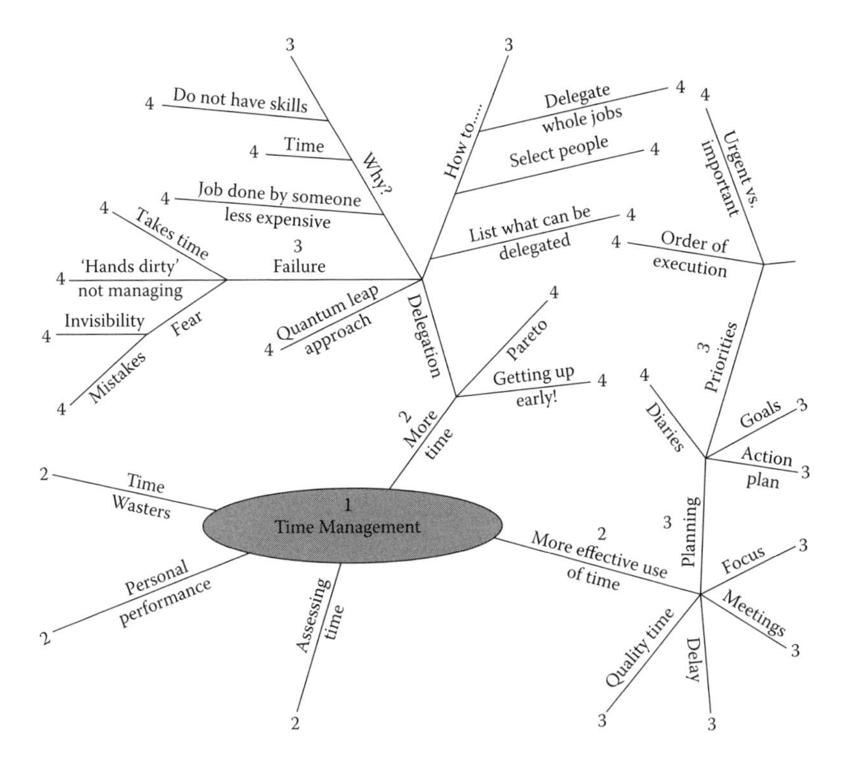

FIGURE 13.18 Mind mapping.

- To separate the vital or critical few items from the trivial many on a large list. This concept of item separation was originated as a part of Pareto analysis, one of the 7-QC tools, which is discussed later in this chapter.

Multivoting begins with each team member receiving the right to vote up to *X* times, where *X* is approximately equal to half the number of items on a list. Then the members vote individually for the items they believe have the highest priority. At this point the team leader or facilitator compiles the total votes given to each item. A tally sheet, one of the 7-QC tools, is well-suited to this task. The team then selects the leading four to six items for discussion and prioritization relative to each other. When the top four to six items can't be established, those having the fewest votes are removed from the list and another vote is conducted.

13.5.21.1 Multivoting Example

Members of a home builder's sales and marketing team attended many meetings at various sales locations around the city. The meetings were not always as productive as they might have been, so the sales managers from the sales offices called a meeting hoping to improve the situation. A brainstorming session produced the following list of concerns:

1. No agenda
2. No clear objective

3. Going off on tangents
4. Extraneous topics
5. Unproductive
6. Time spent on travel
7. Money spent on travel
8. Too much "dog and pony"
9. Problems not mentioned
10. Unclear charts
11. Few meaningful metrics
12. Trouble calling home office
13. No parking
14. No administrative support

To reduce this list to a manageable size, each group member was given seven votes (half of the total of 14 items). The problems received votes as noted on the following tally sheet:

I	1.	No agenda
IIII	2.	No clear objective
II	3.	Going off on tangents
I	4.	Extraneous topics
II	5.	Unproductive
IIIII	6.	Time spent on travel
IIIII	7.	Money spent on travel
IIIII	8.	Too much "dog and pony"
II	9.	Problems not mentioned
IIIIII	10.	Unclear charts
IIII	11.	Few meaningful metrics
II	12.	Trouble calling home office
	13.	No parking
I	14.	No administrative support

As a result of the vote, the group chose to focus on problems 2, 6, 7, 8, 10, and 11.

13.5.22 Nominal Group Technique

The nominal group technique (NGT) is a variation on brainstorming, which was discussed earlier in this chapter. Its purpose is the same, but the methodology is somewhat different. In group problem solving, sometimes there will be subtle pressure toward conformity (e.g., "If you don't agree with the group, you're not really a team player."). Sometimes, status can impact the group effort; for example, senior or more experienced members of a team are assumed to be more knowledgeable, thus making the less senior members reluctant to participate or to venture forth with completely different ideas. It's also easy for groups to go off on tangents, getting too caught up with one idea or train of thought; members are sometimes unwilling to say something that sounds completely off the wall from what other team members were just discussing. Likewise, the group can be too narrowly focused.

NGT begins with all the team members writing down their ideas on sheets of paper without any discussion. After everyone has generated a list, the facilitator uses a round-robin method of listing each person's ideas on flip charts, taking one idea from each person in succession until all ideas have been listed. There is no discussion during this phase either. Only when all ideas have been written down and posted around the room does the group begin discussion, asking for clarification of each idea. The person who listed the idea should respond to any questions about the idea, but others are encouraged to join in. Sometimes, ideas can be combined if the group agrees that they are essentially the same or closely related.

The next phase of the NGT is to narrow the list to a manageable number, that is, to prioritize each idea. This can be accomplished in several ways. One of the best is multivoting, discussed earlier in this chapter. Once the list of items has been prioritized, the group can begin to develop an action plan for CI, discussed throughout this book.

13.5.22.1 Nominal Group Technique Example

The following office problems were identified in a brainstorming session:

A. Ineffective organizational structure
B. Poor communications outside the office
C. Lack of training
D. Poor communications within the office
E. Unclear mission and objectives
F. Poor distribution of office mail
G. Lack of feedback on reports on management

Each group member wrote the letters A through G on a piece of paper and prioritized each problem from 1 to 7 (lowest to highest), using each number only once. Then the results were summarized and discussed.

13.5.23 Paired Comparisons

Paired comparisons is a step-by-step method for rank ordering a small list of items (usually no more than 10) in priority order. This method was created to help a team prioritize its list of items as well as to help team members make decisions by consensus. Ranking is a useful and important tool since multiple ideas, opportunities, and challenges always need to be prioritized from most to least desirable or costly.

One of the easiest and fastest techniques available for ranking is paired comparisons. To begin, a team leader or a group facilitator prints (or clearly writes) the identity of each item on its own slip of paper, preferably paper with a light adhesive backing. Then all the paper slips are placed on a vertical surface such as a whiteboard, window, or wall where all the participants can easily view the ranking activity. Now the team leader or group facilitator starts the paired comparisons process by arbitrarily selecting any two slips of paper and placing them adjacent to each other on the same vertical surface in proximity to the remaining slips. The leader or facilitator then asks the team or group, "Which one of these two items is more (or less) important (or costly)?" Depending on the consensus, the slips are then placed one above the other with the better choice located above the lesser choice.

When this ranking is completed, a third slip is selected and compared with the better choice of the previous ranking. Depending on the resulting consensus, the third slip may be placed above or below the better choice of the previous paired comparison. If it's above, this phase is complete. If it's below, then it's time to compare the third slip with the lesser choice of the first paired comparison. Again, depending on the resulting consensus, the third slip may be placed above or below the lesser slip of the initial paired comparison. This completes the second phase.

Next, arbitrarily select the fourth slip, and repeat the placement process. When all of the slips in the original collection have been placed in the ranked sequence of slips, the process is complete. The top slip is the most important/least costly/best choice. The bottom slip is the least important/most costly/worst choice. Figure 13.19 illustrates the paired comparisons process.

13.5.24 Pairwise Ranking

Pairwise ranking is a structured method for ranking a small list of items in priority order. It can help a team to prioritize a small list as well as to make decisions by consensus.

To begin a pairwise ranking, a pairwise matrix is constructed. Each box in the matrix represents the intersection (or pairing) of two items. When the list has five items, the pairwise matrix will look like Figure 13.20a, with the top box representing

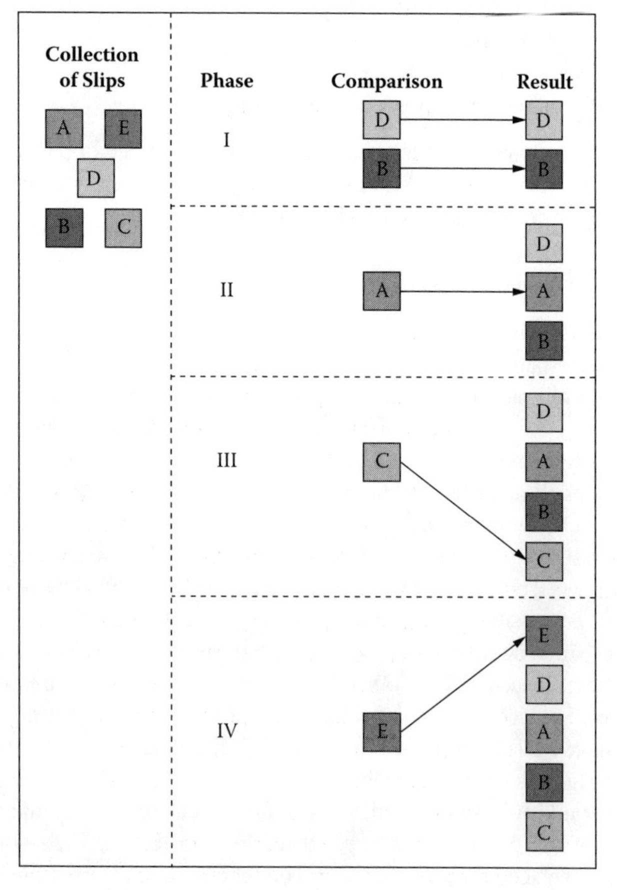

FIGURE 13.19 Paired comparisons.

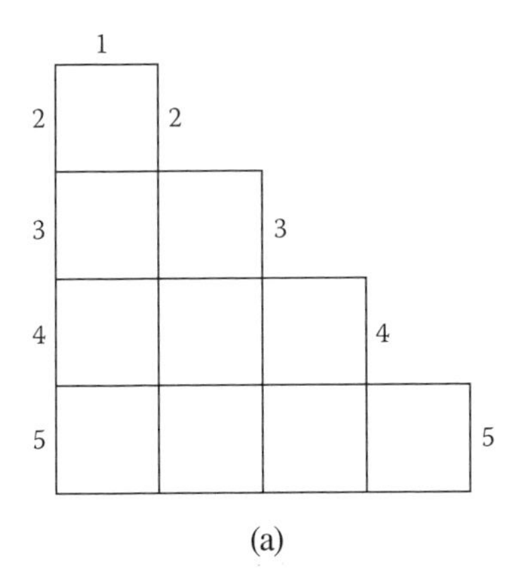

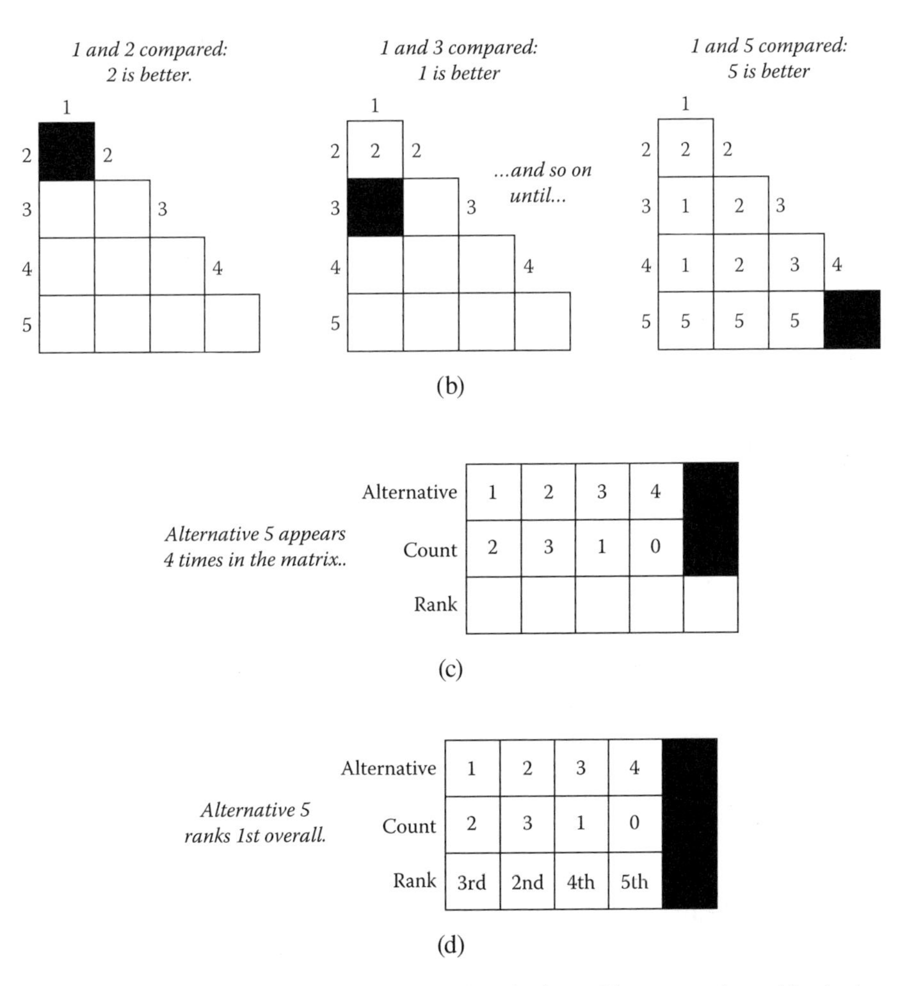

FIGURE 13.20 (a) Pairwise ranking: template; (b) pairwise ranking: comparisons; (c) pairwise ranking: counts; (d) pairwise ranking: final rankings.

Idea 1 paired with Idea 2. The team begins the process using consensus to determine which of two ideas is preferred for each pairing. Then for each pair, the number of the preferable idea is written in the appropriate box. This process is repeated until the matrix is completed. At this point, the team counts the number of times each alternative appears in the matrix.

Finally, the alternatives are ranked according to the total number of times they appear in the matrix. To break a tie, where two ideas appear the same number of times, the box in which those two ideas are compared is examined. The idea appearing in that box receives the higher ranking.

13.5.25 Pareto Analysis

A Pareto diagram is a special form of bar chart used by cross-functional teams to identify and prioritize certain areas of concern (i.e., a change from the status quo). This change could be either a leading process-related problem such as a particular type of defect or complaint, or an unexplained improvement in a product lot or a process output. A Pareto analysis directs a team's attention to the most frequently occurring or most costly area of concern.

There are three uses and types of Pareto analysis. The *basic* Pareto diagram identifies the "critical or vital few" causes that account for the majority of occurrences of the effect under study. The *comparative* Pareto diagram is time oriented and focuses on the status of a system, process, or situation on a before-and-after basis, that is, what changes have taken place as a result of an intervention or corrective action. The *weighted* Pareto diagram provides a measure of significance to factors that may not appear at first to be significant (e.g., cost, time, criticality).

Vilfredo Pareto was an economist who lived in Italy during the late nineteenth and early twentieth centuries. His writings on European economics included a description of maldistribution of wealth within European societies of his time. According to Pareto, this was caused primarily by the excessive land holdings of the aristocracy.

In his early writings, Dr. Joseph M. Juran included a description of the Pareto principle. According to Juran, the greatest payoff comes from focusing on the critical or vital few and by temporarily setting aside work on the trivial many. He advocated use of Pareto analysis to make it simple to select the most important areas of concern. The Pareto principle is often succinctly described as the 80–20 rule. An example would be that 80% of the problems associated with a given situation are a result of 20% of its potential problem sources.

Pareto diagrams are primarily used to help analyze attribute data, whereas histograms are primarily employed to help understand variable data. Figure 13.21a provides a Pareto diagram of nonconforming items, whereas Figure 13.21b is a comparison of Pareto diagrams before and after improvements.

13.5.26 Prioritization Matrix

When a cross-functional team needs to select the "best" way (e.g., method, tool, place, approach) to achieve a specific goal or a task, the prioritization matrix should be the tool of choice. This tool is designed to help guide a team through a logical

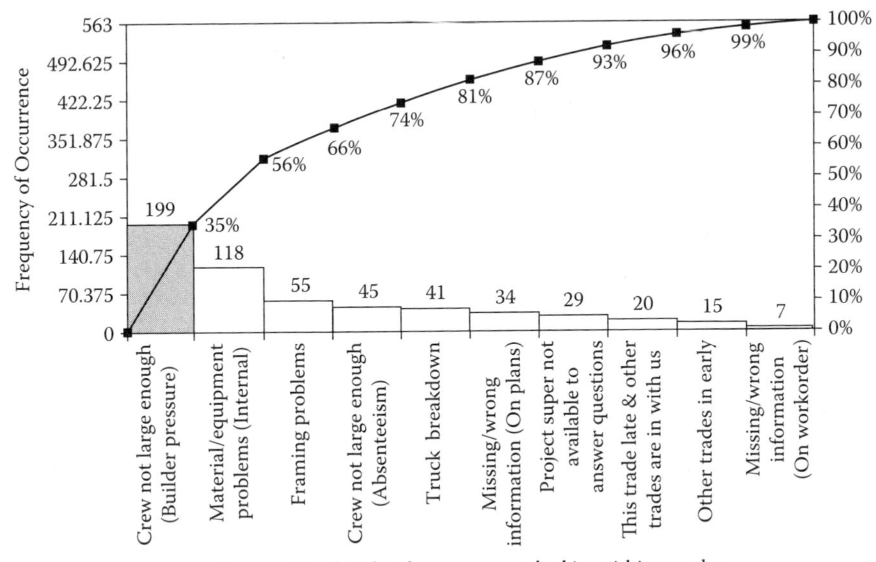

(a)

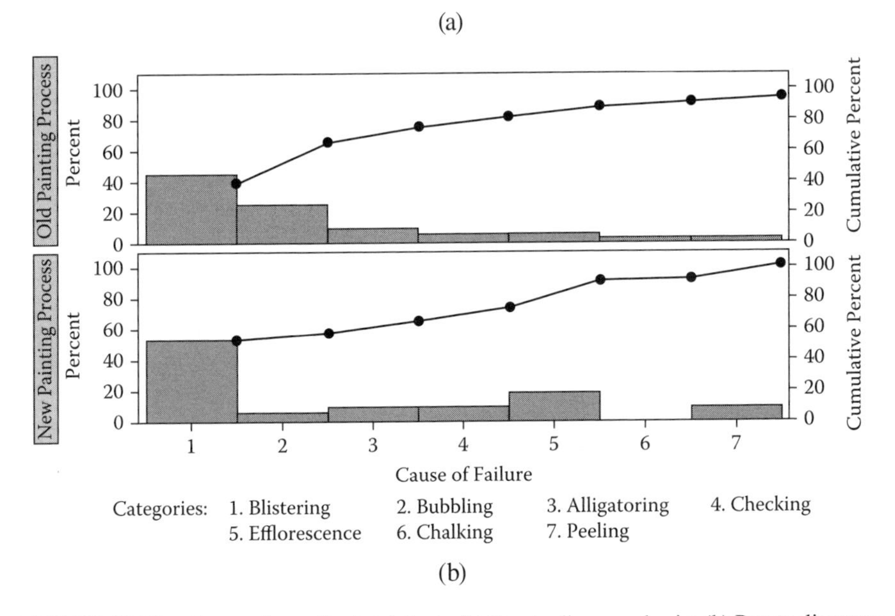

(b)

FIGURE 13.21 Comparison of paint defects: (a) Pareto diagram: basic; (b) Pareto diagram: before-and-after comparison of IC failures.

reduction of available options or choices to identify the "best" one. The reduction process is based on a set of weighted criteria established by the team.

The prioritization matrix uses a simple, two-dimensional matrix format. The row headings consist of the various options being considered. The column headings are the criteria to be used for evaluation of the options. After identifying the pertinent criteria, the team agrees on how the criteria should be weighted. This weighting indicates how each criterion ranks in importance relative to all the other criteria.

The prioritization matrix is an innovation based on the results of combining two other 7-MP tools (i.e., the tree diagram and the L-shaped matrix). Use of the prioritization matrix begins with the selection of two sets of data; each set has been organized with a tree diagram. The outputs of the tree diagrams are used to develop both dimensions of the L-shaped matrix, which then becomes the prioritization matrix. Figure 13.22 demonstrates a prioritization matrix.

13.5.27 Process Decision Program Chart

The process decision program chart (PDPC) assists cross-functional teams in anticipating events and in developing preventive measures, or countermeasures, for undesired occurrences. It is typically used when a project or task is unique, the situation is complex, and the price of a potential failure is unacceptable.

The PDPC is similar in appearance and, to some extent, an application of the tree diagram. It is designed to lead a cross-functional team through the identification of tasks and paths necessary to achieve a goal as well as the associated subgoals. The PDPC then leads the team to respond to the questions, "What could go wrong?" and "What unexpected events could occur?" Next, this tool leads the team in the development of countermeasures. The PDPC provides effective contingency planning

Options	Cost (0.26)	Time (0.14)	Resistance to Change (0.01)	Impact on Problem (0.59)	Row Total (1.000)	Rank Order
A	(0.26) (0.31) = 0.081	(0.14) (0.22) = 0.031	(0.01) (0.11) = 0.001	(0.59) (0.29) = 0.171	0.284	1
B	(0.26) (0.12) = 0.031	(0.14) (0.23) = 0.032	(0.01) (0.37) = 0.004	(0.59) (0.19) = 0.112	0.179	3
C	(0.26) (0.12) = 0.031	(0.14) (0.22) = 0.031	(0.01) (0.02) = 0.000	(0.59) (0.27) = 0.159	0.221	2
D	(0.26) (0.33) = 0.086	(0.14) (0.19) = 0.027	(0.01) (0.29) = 0.003	(0.59) (0.04) = 0.024	0.140	5
E	(0.26) (0.12) = 0.030	(0.14) (0.14) = 0.020	(0.01) (0.21) = 0.002	(0.59) (0.21) = 0.124	0.176	4
				Grand Total	1.000	

FIGURE 13.22 Prioritization matrix.

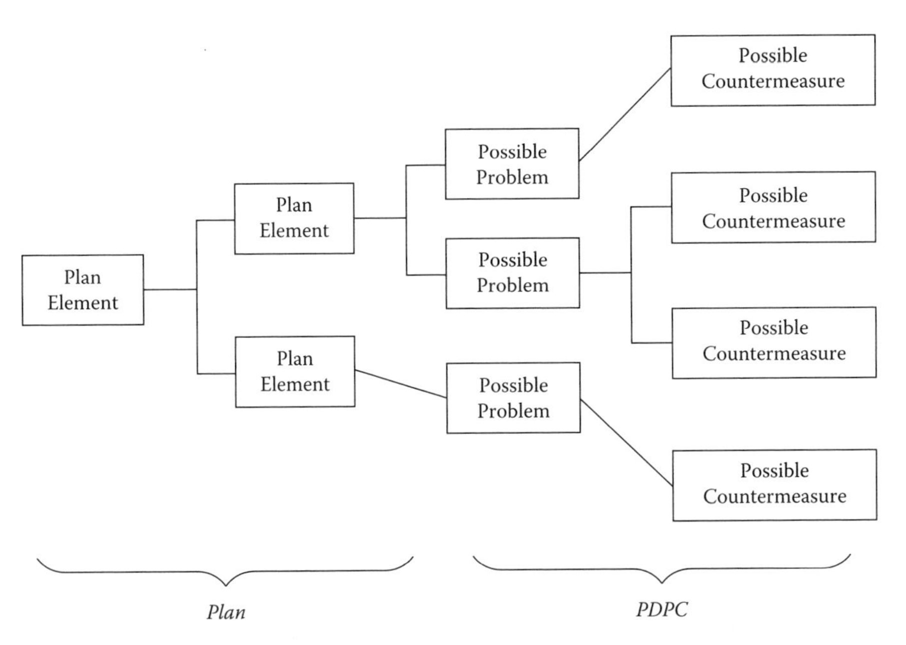

FIGURE 13.23 Process decision program chart (PDPC).

by mapping out all the conceivable events and then by directing attention to where appropriate countermeasures require development.

The PDPC is an effective tool for evaluating a process or activity. It maps out the conceivable activities or actions as well as the associated contingencies in a methodical manner that is easy to explain. It is especially powerful when used in conjunction with other 7-MP tools. Figure 13.23 provides an example of a PDPC.

13.5.28 Process Flowchart versus Process Map

The concepts of process and system are discussed elsewhere in this book. This section reviews what can be done to better understand these concepts. There are a number of different ways to analyze a process. The most common and one of the most useful forms is a graphic tool known as a process flowchart. This chart is a series of geometric figures—rectangles, diamonds, and circles or various other shapes—arranged typically from left to right and from top to bottom, connected by lines with arrowheads to show the flow of activity from the beginning to the end of the process.

13.5.28.1 Process Flowchart

When a process is being created or an existing process is being analyzed, it is useful to create a process flowchart so that everyone involved (i.e., all of the process stakeholders) can see exactly what is supposed to happen from beginning to end without having to try to imagine it. Each of us, in our own minds, may have a picture, a graphical image of what a process flow looks like. However, it may be different from the way others picture it, so one way we can be certain we share a common perspective or outlook of a process is to graph it as a process flowchart—a linear or one-dimensional, graphical portrayal of a process including points of inspection and test. It clarifies the interrelationships that exist between the tasks or steps and is designed to ensure a clear and common vision of a process. *One-dimensional* is used here to distinguish it from the two-dimensional graphic known as a process map, discussed shortly. Figure 13.24a is an example process flowchart, and Figure 13.24b is an actual process flowchart for concrete footing and stem.

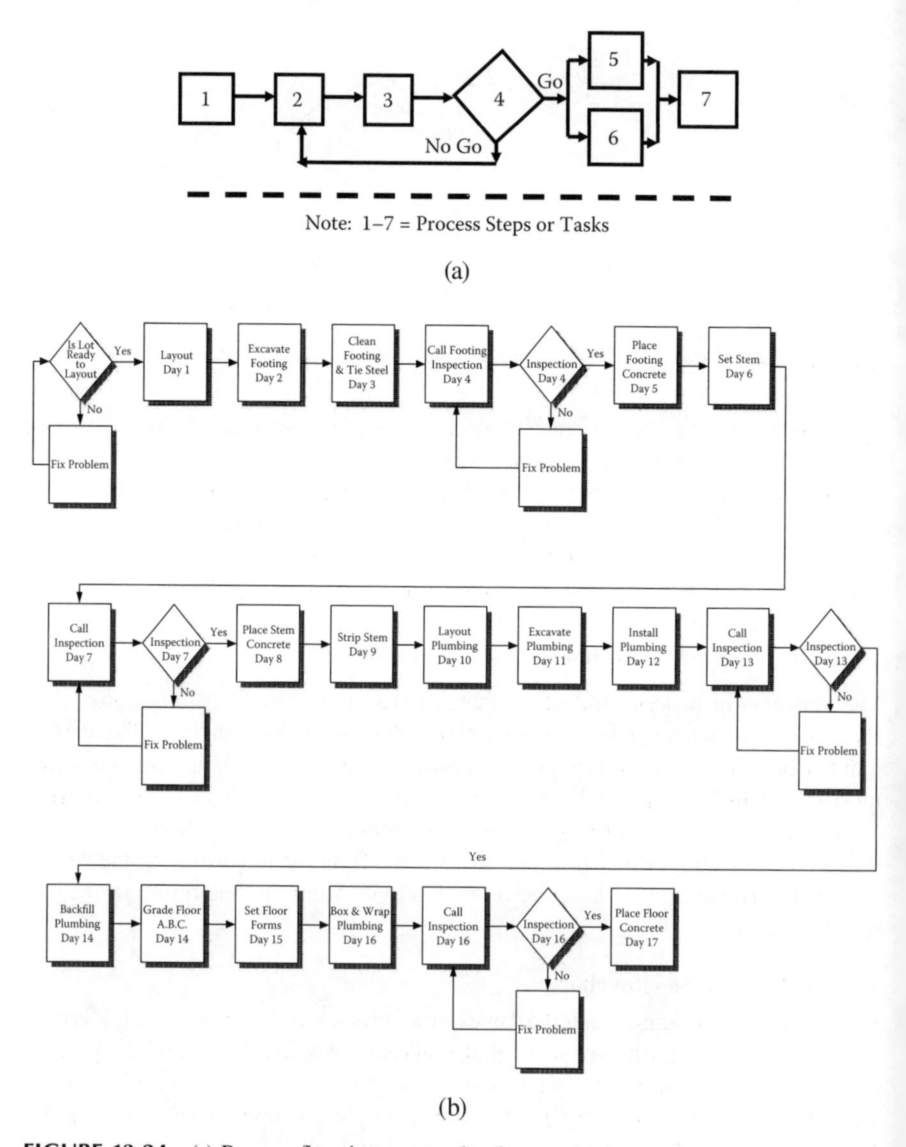

FIGURE 13.24 (a) Process flowchart: example; (b) process flowchart: example—concrete footing and stem.

In addition, a process flowchart distinguishes between value-added and nonvalue-added tasks or steps. This attribute facilitates the reduction of a process cycle time, which subsequently leads to productivity improvement as well as the reduction of defects and variation and, ultimately, the hidden factory (that portion of a company's equipment, tools, machines, personnel, vehicles, supplies, etc. used to correct mistakes, defects and other errors of omission and commision).

First, let's review the steps in the creation of a process flowchart. Traditionally, people have created process flowcharts from the first step to the last step. We don't, and for good reason: When people are developing flowcharts, they are looking at their processes in the same way they look at them every single day. This results in a high potential for missing something important. What we suggest people do as they are brought together to create a process flowchart is to start with the end in mind, a term understood by everyone familiar with Stephen Covey's *7 Habits of Highly Effective People.*

We begin by defining the last step or the output of the process and then start asking the question sequentially, "What has to happen just before that?" If we know we have a specific output or step, we ask what predecessor event or events must take place to satisfy all the needs so that the step we are focused on can take place. Thus, we work backward from the last step to the first step and keep going until someone says, "That's where this whole thing begins." Now we have defined, from the end to the beginning, the process, graphed as a process flowchart.

Some people might question why it should be done this way. A very effective analogy is this: Suppose you were asked to recite the alphabet; you would say A, B, C, D, E, F, G … in order and without thinking, because you have done it hundreds, perhaps thousands of times. But if you were asked to recite the alphabet backward, you would probably say Z and then have to stop and think what happens before that (i.e., what letter precedes Z). What most people do, we have discovered, is first to do it forward to find out what the letter is and then come back and say that the letter before Z is this and the letter before that is this and so on. Working the alphabet backward makes people look at the alphabet in a way that they have never looked at it before, noticing the interrelationships between the predecessor and the successor letters (events).

The same psychology of working backward applies in dealing with processes, whether we are dealing with a process of building a home, working with accounts payable, developing a flowchart, or understanding a process as it relates to training. Establishing the process flowchart from the last step to the first step is a very strong and powerful way to help people understand what their processes really look like.

13.5.28.2 Process Map

Once a process flowchart has been created and the PIT is satisfied that it truly reflects the order in which the events take place regarding predecessor and successor events, the next step is to create a process map. This is a two-dimensional, graphical portrayal of a process that incorporates the sequence of process steps with the specific trade, location, area, or department where each step occurs. Neither the process flowchart nor the process map should be any more detailed than necessary to ensure its practical use.

Beyond the provision of the clear and common vision, the greatest advantage of a process map, when it is annotated with product or service flow rates, is in its use to analyze bottlenecks, locate sources of delay, and accumulate waste.

Earlier it was noted that a process map is created in two dimensions. The same steps that were identified in developing the process flowchart are used, except now instead of just having the flow go from left to right, the people, trades, positions, departments, or functions involved in the process are first listed vertically down the left side of the process map from top to bottom.

For example, the list might be departments A, B, and C; persons X, Y, and Z; or departments such as purchasing, operations, and accounting. Then, we take the rectangles created in our process flowchart and associate them with the various functional areas, departments, or persons listed down the left side of the process map. What appears is a series of rectangles being built across the map from left to right and simultaneously moving up and down the vertical axis created on the left side of our map. In so doing what we see is very much like a saw-tooth effect with blocks going up, down, and across. Now we have a view of the handoffs from one person to another, one function to another, or one trade to another. We can see where queues (waiting lines) are being built and the potential for excess work in process (WIP) is being created among the various areas of responsibility that are listed down the left side.

This provides a clear, visual picture of some of the actions or events we might want to consider for reordering or modifying the various process steps. The objective is to minimize the total number of handoffs that are a part of a process, recognizing that every time there is a handoff, there is the strong potential for an error, an oversight, something left out, a buildup of a queue, the creation of a bottleneck, or the like. Figure 13.25 presents an example process map for the same process portrayed in the process flowchart contained in Figure 13.24.

While creating a process map, tremendous insights are gained into what can be done to continuously improve our processes. Remember: the order of the steps may have been absolutely vital at one time; but with changes in technology, people, and responsibilities, what is being done today may no longer be valid, so we need to periodically assess or review our processes. The use of a process map is an excellent way to do that.

Now, in addition to looking at the process flowchart and process map in terms of the sequence and the handoffs, we can also use these graphics to assess cycle time and value-added versus nonvalue-added events or steps in a process. An effective

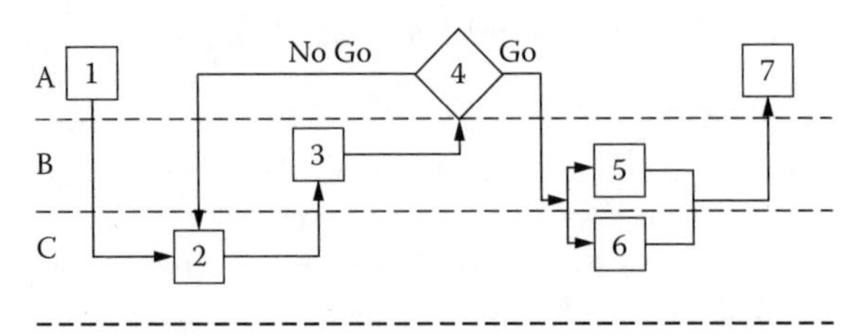

FIGURE 13.25 Process map.
Note: 1–7 = Process Steps or Tasks
A–C = Process Stakeholders (Persons or Departments)

technique is to ask everyone in a room to assess the overall cycle time duration of the process that was just evaluated using a process map or process flowchart. Does it take three hours, five days, 10 weeks? When an agreement is reached of six to eight hours or six to eight weeks—or whatever the final range may be—then each individual step is evaluated. This is accomplished by asking how long each step takes and using the median value as our best choice. When all the individual steps have been estimated, the grand total of all the individual step estimates are summed and then compared with the estimate that the group has already made of the overall process.

What is frequently found is that the sum of the individual steps is only 20 to 30% of the overall total. That very quickly presents an image of a great deal of lost and wasted time—costly time that could be used for other important purposes instead of being thrown away. If, for example, a process is estimated to take six weeks but the sum of the individual components takes a week and a half, it is apparent we have some time that can be saved. Now, what needs to be done? Where are the barriers, the bottlenecks in the process that can be studied; where can, for example, our departments share responsibility? Simultaneously sharing responsibilities instead of passing responsibilities back and forth several times to do some little job is an effective way of reducing cycle time. Steps can be eliminated, and days upon days can be banked for use in more important projects.

13.5.29 Progress Center

A progress center, or a process progress center as it is sometimes known, facilitates continuous improvement of a process by bringing together pertinent diagnostics such as the 7-QC and the 7-SUPP tools.

Tremendous synergy is achieved by collecting these tools in a single, easy-to-access location in proximity to a process. Process personnel, including process operators, engineers, and supervisors, use a progress center to regularly check the status of a process. This is accomplished through simultaneous review and analysis of all the charts, diagrams, graphs, and logs. The review and analysis include a check for consistency between each of the various progress center components to ensure that figures noted on one component are in harmony with the others. Figure 13.26 offers a progress center template.

13.5.30 Project Status/Power Curve (PS/PC)

At the outset of virtually every CI project, each PIT leader is expected to work with his or her team members to plan out the project schedule and, subsequently, to convert this project schedule into a PS/PC. The initial and all follow-up versions of the PS/PC are submitted to the continuous improvement manager (CIM). The CIM is responsible for the monthly review of all the PS/PCs and for presenting a summary status of all CI projects to the home builder's executive committee.

To prepare a PS/PC effectively and efficiently, the PIT must take a number of preliminary steps. This is accomplished with the leadership of the team leader and the guidance of the team advisor (a Six Sigma/Black Belt or an external consultant). The remainder of this section describes each step using an ongoing example for clarification.

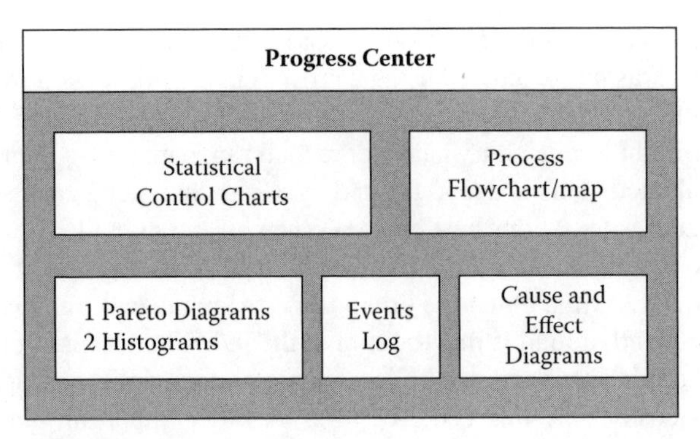

FIGURE 13.26 Progress center.
1 For attribute (discrete) data
2 For variable (continuous) data

13.5.30.1 Scope the Complete Project

Estimate the duration of the project in the most appropriate time units (e.g., seven weeks, three months). This will require some discussion with the team members who will be doing their regular jobs while they are involved in collecting and analyzing data as well as participating in team meetings. For our example, let the expected project duration be 20 weeks.

13.5.30.2 Scope the Individual Project Phases

Using the CI problem-solving methodology described in Chapter 4 (page 26), each project will have five phases (B.U.I.L.D.): B = Begin with process pain; U = Understand Continuous Improvement; I = Improve the Process; L = Learn the Lessons; and D = Deploy the Learning Step. Recognizing that some phases will take more or less time than other phases, estimate the duration of each phase. The sum of the estimated time durations for the five phases must equal the duration of the project estimated in Step 1. For our ongoing example, let the first phase (Define) be one week, the second phase (Measure) be nine weeks, the third phase (Analyze) be five weeks, the fourth phase (Improve) be four weeks, and the fifth phase (Control) be one week. Thus, B.U.I.L.D. sums to 20 weeks.

13.5.30.3 Convert Individual Project Phase Durations into Percentages

Set up a table with the information from Step 2, as follows:

SS Phase	Duration
Define	1 week
Measure	9 weeks
Analyze	5 weeks
Improve	4 weeks
Control	1 week
Project duration = 20 weeks	

Now convert each of the phase durations into whatever percentage of the entire project it represents by dividing it by the overall project duration, as follows:

SS Phase	Duration	Percentage
Define	1 week	1/20 = 5% for Define Phase
Measure	9 weeks	9/20 = 45% for Measure Phase
Analyze	5 weeks	5/20 = 25% for Analyze Phase
Improve	4 weeks	4/20 = 20% for Improve Phase
Control	1 week	1/20 = 5% for Control Phase
Project duration = 20 weeks		100%

These percentages will be used to establish the PS/PC baseline.

13.5.30.4 Create the PS/PC Baseline

There are three parts to this step. The first part is to establish the horizontal axis, "Cumulative Weeks (Cumulative % of Time)." The second part is to establish the vertical axis, "% Project Completed." The third part is to draw in the PS/PC baseline. The PS/PC baseline is a line of points where the percent on the horizontal and vertical axes are equal. This concept is demonstrated in Figure 13.27a. Let's look at the results of the first part (i.e., establishing the horizontal axis). In this ongoing

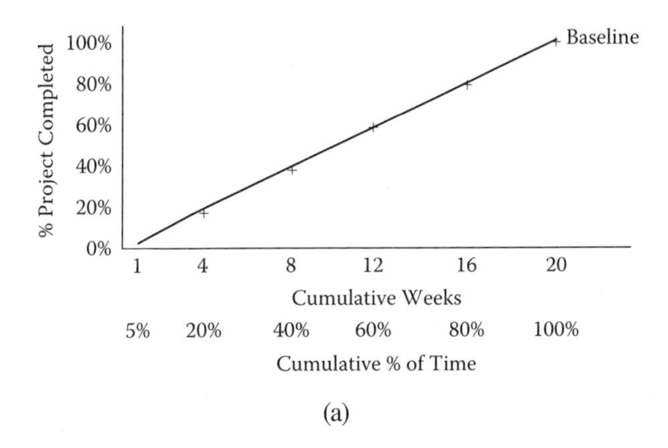

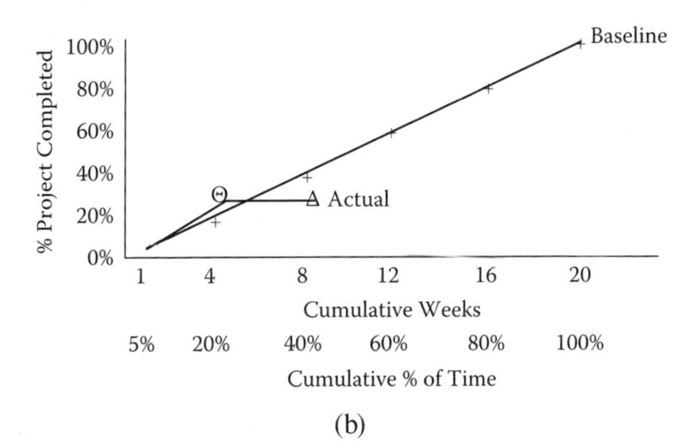

FIGURE 13.27 (a) Project status/power curve (PS/PC): baseline; (b) project status/power curve (PS/PC): simple example; (c) project status/power curve (PS/PC): complex example.

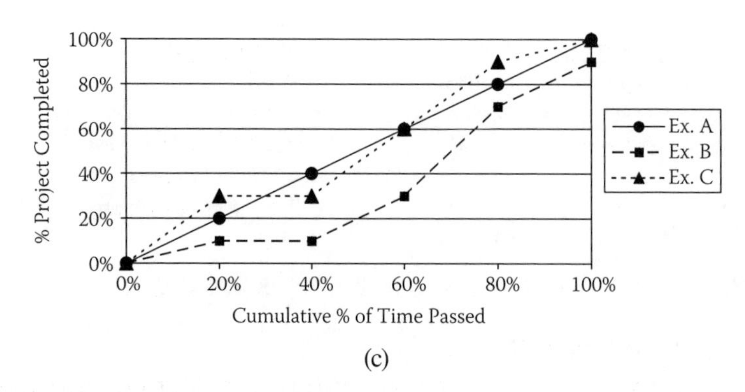

FIGURE 13.27 *(Continued).*

example, the first week is 5% of the 20-week project duration (1/20 = 5%), the first four weeks is 20% of the 20-week project duration (4/20 = 20%), the first eight weeks is 40% of the 20-week project duration (8/20 = 40%), and so on.

13.5.30.5 Plotting Project Status against the Project Baseline

As the team makes progress in working toward its goals as stated in its project charter, the team leader in consultation with his or her team members must estimate the extent to which its goals have been achieved. Let's suppose that when four weeks (20% of the original estimate of the total project duration) from the project kickoff have passed, it's time to submit a PS/PC to the CIM. Together, the team leader and the team members along with the team advisor estimate what percentage of the overall project has been completed so far. In our ongoing example, let's say that the team estimates it has completed 30% of the project.

Our first status point on our PS/PC would appear as Θ (four weeks on the horizontal axis and 30% on the vertical axis).

Going a bit farther in time to eight weeks (40% of the original estimate of the project duration), it's time to submit a second report, and the team decides that it has fallen behind (the power curve baseline) since nothing has really been accomplished since the first report was submitted. Thus, the second status point on the PS/PC would appear as **Δ** (eight weeks on the horizontal axis and 30% on the vertical axis). This sequence of events continues until the project has been completed. A team may complete its work ahead of schedule (the PS/PC baseline), on schedule, or behind schedule.

In any case the use of the PS/PC to verify what has been achieved and reported on a regular basis will help keep the team and the home builder's executive committee aware of the good, the bad, and the ugly things that happen, as pictured in Figure 13.27b.

13.5.30.6 Another Example

Suppose we have a project that is estimated to take about six months to complete. Let's call it 25 weeks to make the example calculations easier. If we stay on schedule, our project status is right on the power curve, as follows:

Time (Weeks)	Percentage of Time Passed	Cumulative Percentage of Time Passed
1	4	4
5	20	20
10	20	40
15	20	60
20	20	80
25	20	100

Now suppose that at each point on the horizontal time line we can estimate how much of the project has actually been completed (the project status). Here are three examples:

	Percentage of Project Completed			
Time (Weeks)	Example A	Example B	Example C	Cumulative Percentage of Time Passed
1	4	4	4	4
5	20	10	30	20
10	40	10	30	40
15	60	30	60	60
20	80	70	90	80
25	100	90	100	100

Graphically, the examples appear as in Figure 13.27c.

Now you have a way to graphically track project status or progress. When you're below the power curve, your project is in trouble. Being above the power curve is even better than being on it. It should be expected that your original estimate of the overall project duration may just as easily be too long or too short. Try to produce your best estimate of the project duration the first time, but don't be surprised if you and your team need to reevaluate as you learn more about your project.

This doesn't mean that teams should feel free to modify their project schedule whenever the team falls behind its power curve baseline or identifies an approach to problem solving that reduces the original project duration. It does mean that when it becomes apparent that the original schedule was not realistic, the team leader, after conferring with his or her team, should request approval by the CIM to appropriately modify the project schedule and, of course, the PS/PC. And don't be shy about asking for advice and assistance from your team's advisor, either an internal or external consultant.

13.5.31 Quantified Force Field Analysis

Force field analysis is used to visually identify and then quantify the relationships of significant and opposing forces that influence the achievement of an objective or completion of a plan.

Force field analysis can be used in the following instances:

- To identify key factors or forces that either promote or hinder desired accomplishments
- To identify improvement opportunities

This deceptively simple technique begins with a statement of the objective or plan below and to the right of the graphic. This is followed by establishing a vertical line that separates positive forces on the left of the line and negative forces on the right. Each force is portrayed as a straight horizontal line with the arrowhead of the positive force lines pointed to the right and touching the vertical line and the arrowhead of the negative force lines pointed to the left and touching the vertical line. Figure 13.28 provides a template for this graphic.

13.5.32 Random Sampling

Random sampling is a mathematically based technique designed to ensure that every unit in a population has an equal opportunity to be drawn from the population and become a part of a sample. This guarantees that, however many samples may be taken from a population, all the samples fairly represent the composition or makeup of the population.

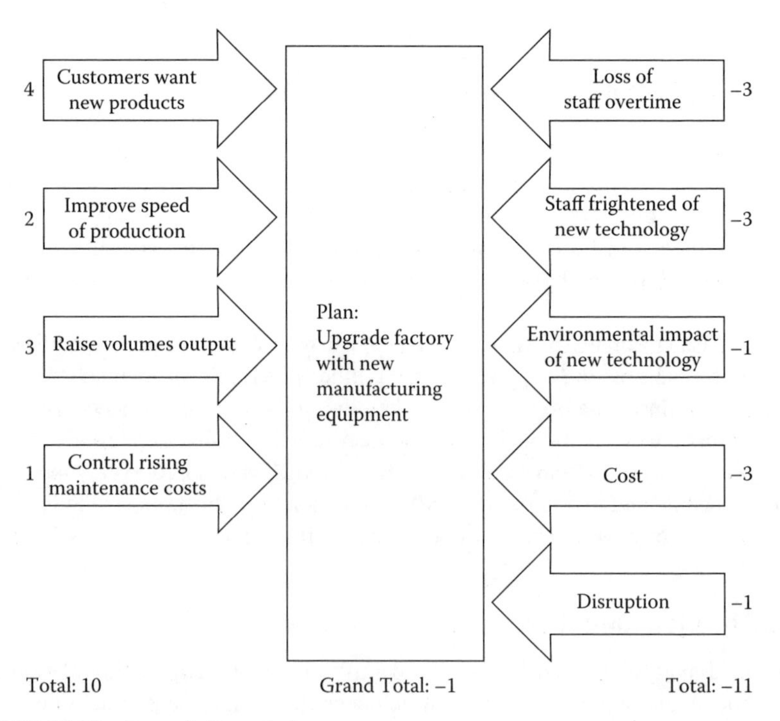

FIGURE 13.28 Force field analysis.

The following are examples of random sampling:

- A sampling of units produced by either a construction or an administrative process to determine if the results of the process are predictable (in statistical control)
- A survey of a builder's employee population to ascertain the extent of approval of an entirely new bonus procedure
- Selection of a sample of homes to be tested to ensure that newly installed windows and doors are leak proof

The application of randomization to statistical sampling ensures that the results of drawing a sample from a population will fairly represent the results that would have been generated had a 100% census been conducted. There are major advantages to random sampling; it requires less time, costs less, provides fewer opportunities for errors in data collection, and involves fewer personnel as both data collectors and possible respondents.

13.5.33 Run Chart

A run chart (also known as a trend chart) is a line chart/graph that permits the study of data over some specified period of time. The data can be either attribute/discrete or continuous/variable. After observations have been made and the associated data have been recorded, the data points are plotted and connected as a run chart. The resulting line allows examination for trends or patterns such as cycles or shifts in magnitude. The chart is typically oriented with time plotted on the x (horizontal) axis and the data plotted on the y (vertical) axis.

The average value of the original input data is calculated and plotted as a horizontal line across the specified time period. The average line is then used to estimate the orientation of the data; that is, is it reasonably constant with minor data dispersion evenly balanced above and below the average line, or is there an upward or downward trend? A median line is sometimes used as the center line because it provides a nonparametric (not dependent on the frequency distribution) estimate of the process centerline.

A run chart is the beginning of a (statistical) control chart. All that needs to be added are the upper and lower statistical control limits, which are calculated based on the original input data.

13.5.34 Samples versus Populations

Whether considering the totality of a process (every step) or an entire inventory (every unit) or a specific portion (randomly selected units) of its products, this focus is referred to as the population. In the world of continuous improvement, we try to discover facts by collecting relevant data and then taking action based on the analysis of the data. Thus, the data are not collected as an end in itself but rather as a means of determining the facts that exist behind the data.

13.5.35 Scatter Analysis

The most basic form of a DOE is a graphical analysis of two types of data. One type of data is treated as being independent, and the other is considered as being dependent, a type of data whose magnitude is somehow related to the value of the first type. When these two types of data are plotted on an x-y coordinate graph, the resulting pattern of bivariate (*x*,y) data points is referred to as a scatter diagram or scatterplot.

Scatter analysis, the interpretation of a diagram or plot, is used to determine the extent to which the two types of data may be related as well as the orientation of the relationship. The extent of the relationship is defined as falling between 0 and 1, whereas the orientation can be either positive or negative.

If, as a result of a scatter analysis, it is apparent that the magnitude of the dependent variable is increasing or decreasing in the same direction as the independent variable, then this is described as being a positive relationship. If the scatter analysis indicates that the dependent variable is decreasing when the independent variable is increasing, or vice versa, then this is a case of a negative relationship. The positive or negative orientation describes the plus or minus value of the slope of a line, which best explains any existing relationship between the two variables. Data correlation patterns for scatter analysis are pictured in Figure 13.29a.

Figure 13.29b is an example of scatter analysis. The bivariate data in the aircraft are accurate but provide no clues as to which airlines are doing well and which ones aren't. The scatter diagram in Figure 13.29c paints a graphic picture of which airlines are in trouble with respect to their costs.

13.5.36 Tally Sheet

A tally sheet provides a systematic method for collecting data. In most cases, it is a form designed to collect specific data. This tool provides a consistent, effective, and

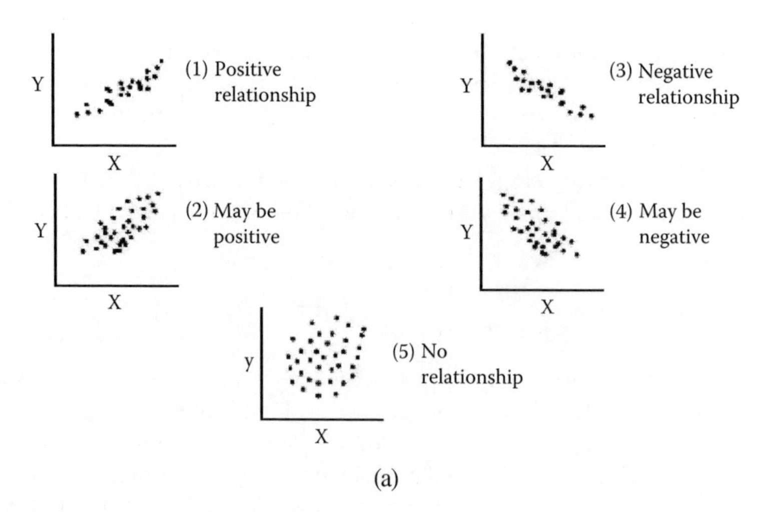

FIGURE 13.29 (a) Scatter analysis: data correlation patterns; (b) scatter analysis: bivariate data; (c) scatter analysis: graphic example.

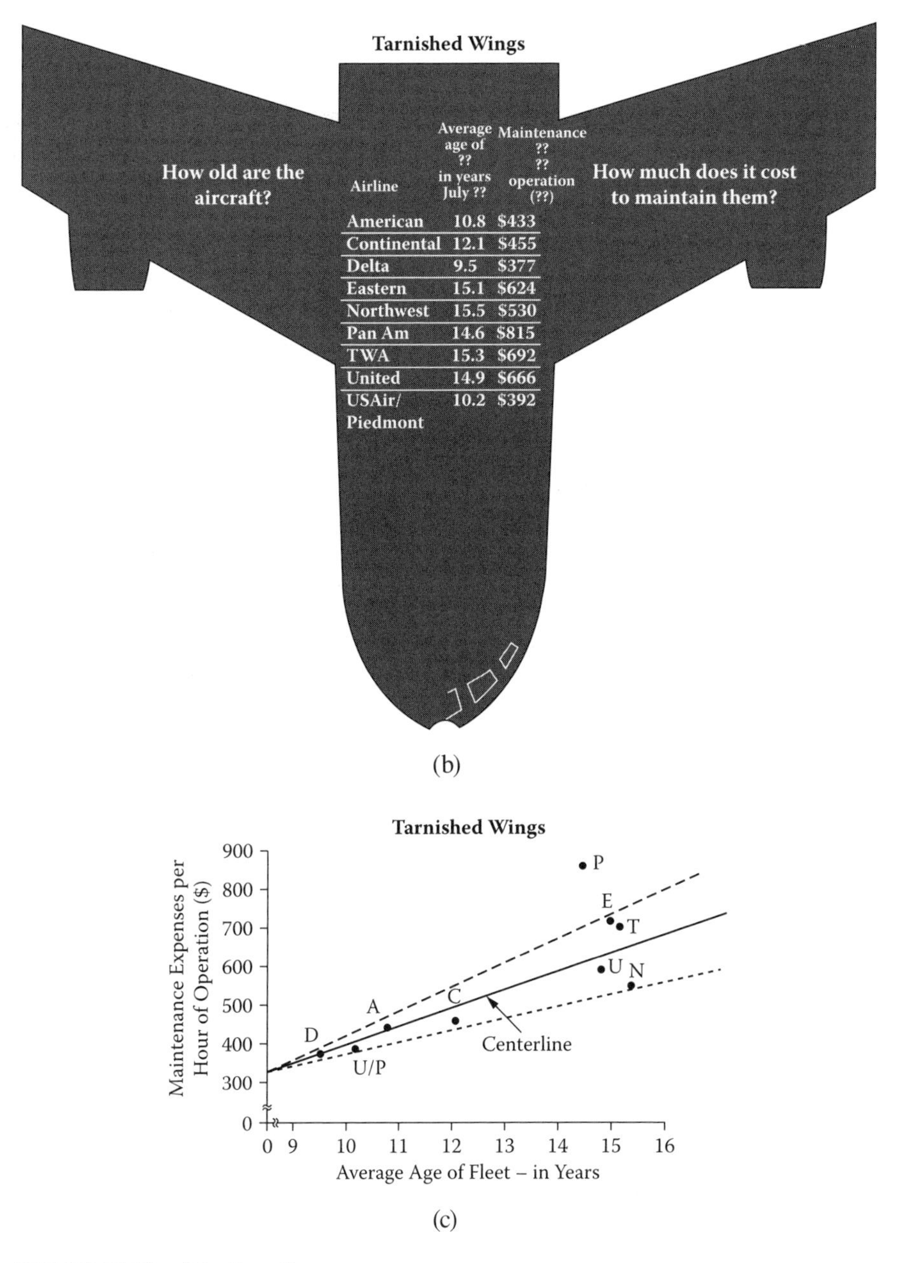

Airline	Average age of ?? in years July ??	Maintenance ?? ?? operation (??)
American	10.8	$433
Continental	12.1	$455
Delta	9.5	$377
Eastern	15.1	$624
Northwest	15.5	$530
Pan Am	14.6	$815
TWA	15.3	$692
United	14.9	$666
USAir/ Piedmont	10.2	$392

(b)

(c)

FIGURE 13.29 *(Continued).*

economical approach to gathering data, organizing it for analysis, and displaying it for preliminary review.

A tally sheet can be used with either attribute or variable data. It sometimes takes the form of a manual data collection form where automated data is either not necessary or not available. It should be user-friendly, that is, designed to minimize the

Events	Day 1	Day 2	Total
A	IIIIII	IIIIIII	13
B	III	II	5
C	IIII	III	7

FIGURE 13.30 Tally sheet.

need for complicated entries. Simple to understand, straightforward tally sheets are key to successful data gathering. Figure 13.30 offers an example tally sheet.

13.5.37 Tree Diagram

The tree diagram has many applications in the TQM/CI arena because it is a systematic tool for determining all the tasks necessary to accomplish a specific goal. It can be used to determine the key factors that are the source of a particular problem or to create a fully developed action plan for a single event or a process. The tree diagram can also be used to decide on the priority for a given action.

This tool uses linear logic to move from a broad statement to successively lower levels of detail. The statement can be generated using affinity analysis, an ID, or brainstorming. A tree diagram is quite useful, therefore, when a task or problem is complex and when it is important to identify all the key elements or subtasks of which the task or problem is composed. Figure 13.31 provides an example application of a tree diagram.

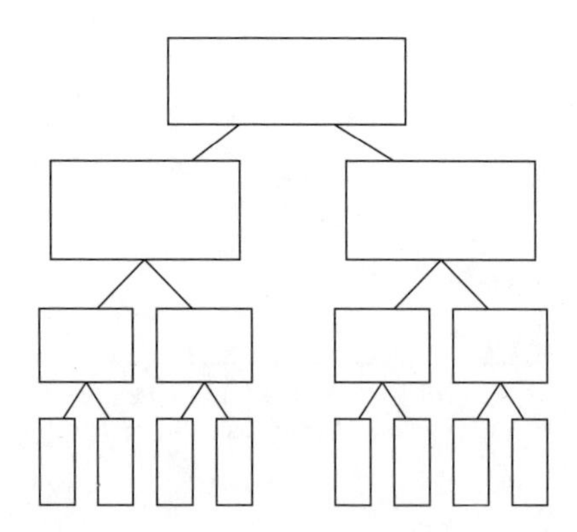

FIGURE 13.31 Tree diagram.

14 Quality Function Deployment

14.1 INTRODUCTION

In the world of business and industry, everyone has customers. Some have only internal customers, some just external customers, and some have both. When a team is working to determine what needs to be accomplished to satisfy or even to delight its customers, then the tool of choice is quality function deployment (QFD), as noted in Figure 14.1.

QFD has been referred to by many names—matrix product planning, decision matrices, and customer-driven design, to name just a few. Whatever it may be called, QFD is a focused methodology for obtaining and listening to the voice of the customer (VOC) and then effectively responding to customers' needs and expectations. One of the major reasons why QFD is so important in residential construction is the all-too-often failure to communicate between home builders and home buyers. Figure 14.2 offers convincing evidence that the voice of the home buyer is not always clearly received by home builders.

Figure 14.3 offers an amusing example of a simple failure to communicate.

First developed in Japan as a form of cause and effect analysis (one of the seven quality control [7-QC] tools) in the late 1960s, QFD was brought to the United States in the early 1980s. It gained its early popularity as a result of numerous successes in the automotive industry.

In QFD, quality is a measure of customer satisfaction with a product or a service. QFD is a structured method that uses the seven management and planning (7-MP) tools to quickly and effectively identify and prioritize customers' expectations:

- Beginning with the initial matrix, commonly termed the *house of quality* (HOQ), the QFD methodology focuses on the most important product or service attributes or qualities. These are called the WHATs (what the customer expects) and are composed of customer MUSTs (unspoken demands), WANTs (spoken desired qualities/attributes/characteristics), and WOWs (unspoken delighters and exciters).
- Once the WHATs and their corresponding customer importance values have been identified, the next step is development of the HOWs (organizational responses to the WHATs) along with the level of difficulty or risk associated with each HOW. After the interrelationships between the WHATs

- **Objective:** Assign home buyer requirements throughout virtual home builder organization
- **Mechanism:** Translate home buyer requirements into appropriate technical requirements for each stage of product design and construction

FIGURE 14.1 Why QFD?

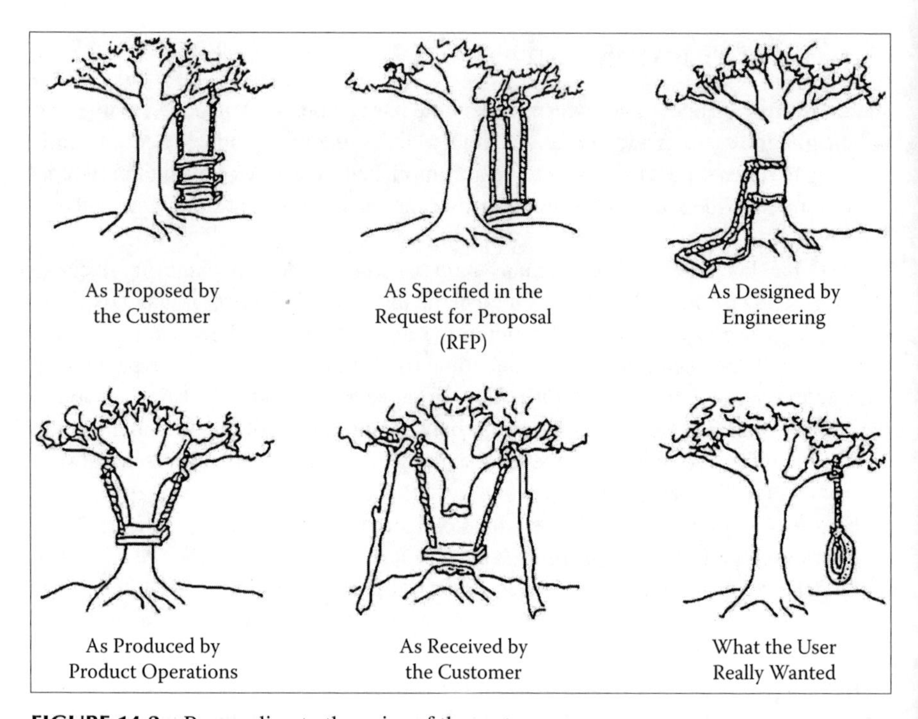

FIGURE 14.2 Responding to the voice of the customer.

- Translation: English to Spanish to English
  - English: Don't Let the Cat out of the Bag
  - Spanish: No Dejes Salir El Gato De La Bolsa
  - English: The Cat Can't Leave the Stock Exchange

- Conclusion:
  - Make Certain Customers Receive Their Quality Requirements
  - Don't Let Poor Communications Translate Customer Quality Requirements into Kitty Litter

FIGURE 14.3 Failure to communicate.

and HOWs have been determined, the HOQ is analyzed to determine the most important HOWs. Then, when the HOWs have been prioritized, QFD deploys responsibility for them to the appropriate organizational function (e.g., construction, operations, customer care, marketing) for action. Thus, QFD is the deployment of customer-driven qualities to the responsible functions of an organization.

There are two major approaches to QFD. The American Supplier Institute (ASI) advocates a four-phase model, while GOAL/QPC uses a 30-matrix model. Depending on what a team's objectives may be, one or the other or a combination of the two models may be the most advantageous approach for the team to apply.

Many QFD practitioners claim that using QFD has enabled them to reduce their product and service development cycle times by as much as 75% with equally impressive improvements in measured customer satisfaction. Improvements of this magnitude exist throughout America in industry, service organizations, and government agencies.

14.2 KANO MODEL

Many "experts" insist that customers don't really know what they want—they have to be told. They're wrong, dead wrong! Home buyers, for example, do know what they want, but unfortunately they're not proficient at describing their needs. When a home builder understands the three types of customer needs and how to reveal them, they are well on their way to understanding their customer's needs as well as, or perhaps better than, their customers do.

The Kano model of customer needs is quite useful in gaining a thorough understanding of customer needs. As portrayed in Figure 14.4, there are two dimensions within the Kano model:

1. Achievement (the horizontal axis), that is, the degree to which execution has been accomplished, runs from the builder "doesn't do it at all" (on the left side of the model) to the builder "does it very well" (on the right side).
2. Customer satisfaction (the vertical axis) goes from "dissatisfaction" with the product or service (at the base of the model) to "satisfaction" with the product or service (at the top of the model).

Dr. Noriaki Kano of Tokyo University isolated and identified three levels of customer needs or expectations (i.e., what it takes to positively impact customer satisfaction). The three levels have been identified as the MUSTs (unspoken needs), the WANTs (spoken needs), and the WOWs (unspoken needs). The remainder of this section describes these three needs.

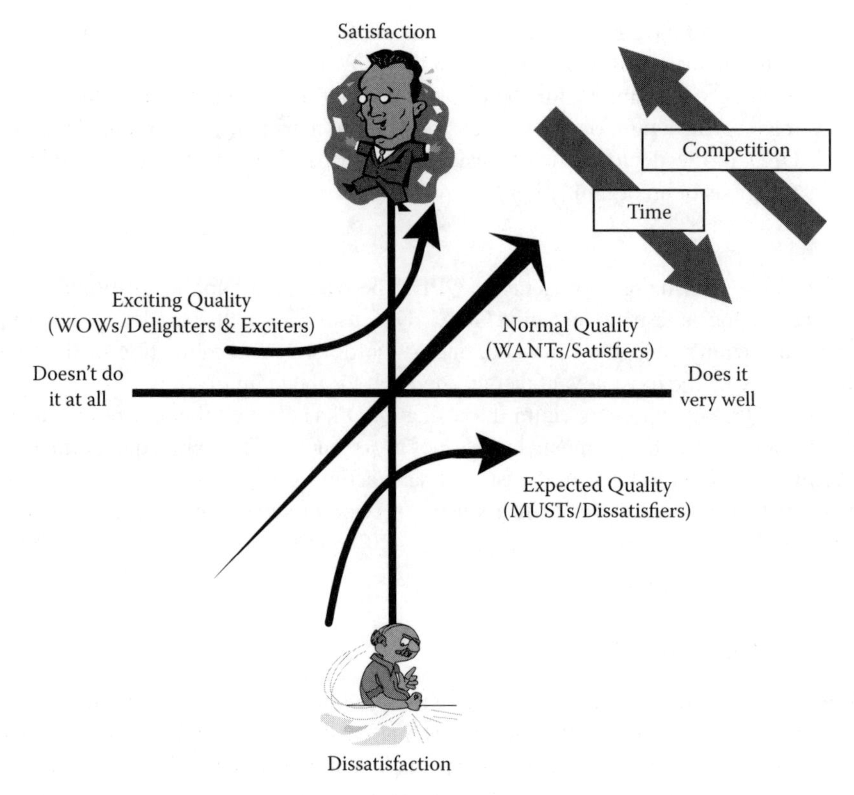

FIGURE 14.4 Kano model of customer expectations.

14.2.1 The MUSTs

Fully satisfying the home buyer at this level simply gets a home builder into the market. The entry-level expectations (i.e., the basic needs) are referred to as the MUST-level qualities, properties, or attributes. These expectations are also known as the *dissatisfiers*, because, by themselves, they are unable to satisfy a home buyer. However, failure to provide these basic expectations will cause dissatisfaction. Examples include features relative to home safety, latest-generation building materials, and the use of branded fixtures and components. The MUSTs include customer assumptions, expected qualities and functions, and other unspoken expectations such as plumbing and air conditioning that work as expected, every time, and a roof that doesn't leak. The must-be-needs include the most basic needs that customers don't mention because they take them for granted. They are dissatisfiers if they are missing, but if they are present they are low on the satisfaction scale since they're expected. Safety, proper operation, and hygiene are examples of this need. Products and services must provide the MUSTs even if the customer did not consciously expect them.

14.2.2 The WANTs

These are the qualities, attributes, and characteristics that keep a home builder in the market. These next-higher-level expectations are also known as the *satisfiers* because they are the ones customers will specify as though from a written or mental list. They can either satisfy or dissatisfy the customer depending on their presence or absence. The WANTs, also referred to as performance or one-dimensional needs, include any spoken home buyer expectations such as Energy-Star appliances and other "green" features as well as many extra-large cabinets and closets. These needs can be identified in formal documents, such as marketing surveys, specifications, and purchasing orders, as well as in less formal sources, such as focus groups and discussions between customers and salespersons. Their contribution to customer satisfaction is directly related to their presence in the delivered product.

14.2.3 The WOWs

These are features and properties that make a home builder a leader in the market. The highest level of customer expectations, as described by Kano, is termed the WOW-level qualities, properties, or attributes. These expectations are also known as excitement needs or more often as the *delighters* or *exciters* because they go well beyond anything the customer might imagine and ask for. Their absence does nothing to hurt a possible home sale, but their presence improves the likelihood of purchase. WOWs excite customers not only to make on-the-spot purchases, but also to return, years later, to make further purchases as well as to refer their friends and relatives. Examples of WOWs include designer chandeliers, bamboo wood floors, and a lifetime guarantee on the roof. These are unspoken ways of knocking the customer's socks off. Attractive needs include features that exceed the requirement and expectations of the customers. These are needs that the producer or service provider can provide; they solve a real problem, but the customers are unaware of them. Delivery of attractive features can significantly increase customer satisfaction, but their absence does not decrease customer satisfaction. Attractive qualities are the product resulting from the talents of the members of the producing organization. These features can delight the customer and become the competitive edge in the market. Remember: your sales and customer service persons must "undercommit and overdeliver," not the other way around.

14.2.4 Changes over Time versus Competition

As demonstrated by the arrows at the top of Figure 14.4, unspoken WOWs become spoken WANTs over time and finally become unspoken MUSTs. Indoor plumbing and bathrooms are immediate examples. The home builder that gets ahead and stays ahead is constantly pulsing its home shoppers to identify the next-generation WOWs. The home builder that wants to succeed over the long haul will never forget that the best WOWs, plenty of WANTs, and all the MUSTs are what it takes to become and remain an industry leader in residential construction or a leader in any business or industry.

The Kano model emphasizes that only close relationships with the customers can provide the deep understanding of their spoken and unspoken needs. This is the only way to improve the customers' satisfaction, to delight the customer, and to achieve an advantage in the competitive market.

The Kano model of customer needs is quite useful in gaining a thorough understanding of customers' needs. The resulting "verbatims" should be translated and transformed using the "Voice of the Customer" (VOC) table, which, subsequently, becomes an excellent input as the WHATs in a QFD HOQ.

14.2.5 How to Create WOWs/Attractive Quality

1. Collect data on current product usage—customer processes and interactions within the "system." Assess customer attitudes about acceptance and application; gather all "goodwill" complaints and categorize them according to type; accumulate "no" data and find out why customers are saying no to your products or services and classify them; observe the actual use of the products and services in the workplace (this is known as "going to the Gemba"); conduct interviews to discover unintended benefits, desires, quality, variation, opinions (i.e., customer intelligence) as used by investigative entities.
2. Extract all the circumstantial issues, no matter how ludicrous or seemingly impossible they seem to be at the time.
3. Confirm in a general nature how these sets of circumstances are related and categorize them by type.
4. Generate ideas from these circumstances about ideas for creating attractive quality.
5. Create an attractive quality survey. (This is a key part of the process; don't rely just on customer comments. Reframe the ideas from the extracted circumstantial comments from the data gathering, formulating them into tangible ideas that customers can actually choose from as to how attractive they might be to them.)
6. Make prioritization decisions based on the results of the survey about which ideas can be generated to create a "pull strategy."
7. Then take the results of the attractive quality creation process to the development stages following the normal course of your development strategy (i.e., product planning, technology development, design, prototyping/testing/evaluation, production, and marketing/promotion/service delivery). Whew!

Dr. Kano provides a clear example of how this works. During the 1970s, Konica camera in Japan discovered what everybody already knew: People who used their cameras had two problems: underexposure and out-of-focus pictures. It was quite clear that these problems had nothing to do with the cameras; they were operator-related problems. Konica could have just created a manual on how to take better pictures and continue to improve the quality of their cameras, but what they did do revolutionized the commercial camera business. In 1974, they built in an automated flash, and in 1977 they created an autofocus camera. Incredibly, they took a problem that was not even their problem and created enhancements to their products, thereby meeting a latent customer need of taking great pictures easily. People flocked to purchase cameras with these capabilities—thus creating a pull strategy!

We all know hundreds of similar cases, but this clearly objectifies the meaning of attractive quality creation and how it develops pull strategies rather than requiring push promoting. When customers "gotta have it," you've invented pull and have made your enterprise attractable.

14.3 THE WHAT–HOW CONCEPT

The basic matrix of QFD is the HOQ, and the core of the HOQ is the WHAT–HOW concept. The WHATs, as noted earlier in this chapter, are the MUSTs, WANTs, and WOWs as described in the Kano model. Figure 14.5 portrays the WHATS (i.e., the voice of the home buyer), whereas Figure 14.6 connects the WHATs to the HOWs (i.e., the organizational responses). The reader should note that for each WHAT, there is at least one HOW that has been identified by the home builder to respond to the WHAT. For example, a WHAT such as extra-large walk-in closets in all bedrooms might

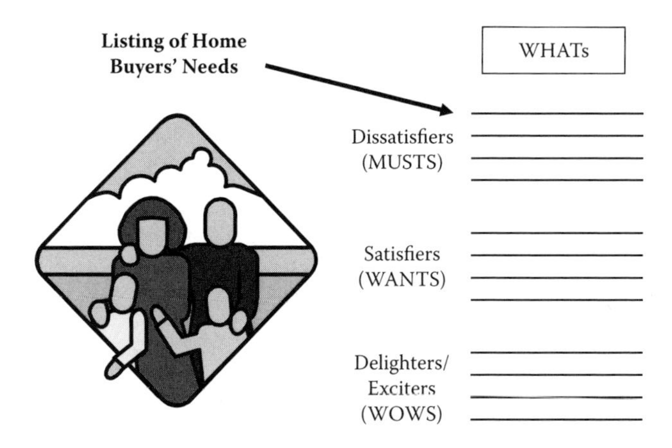

FIGURE 14.5 Voice of the home buyer (WHATs).

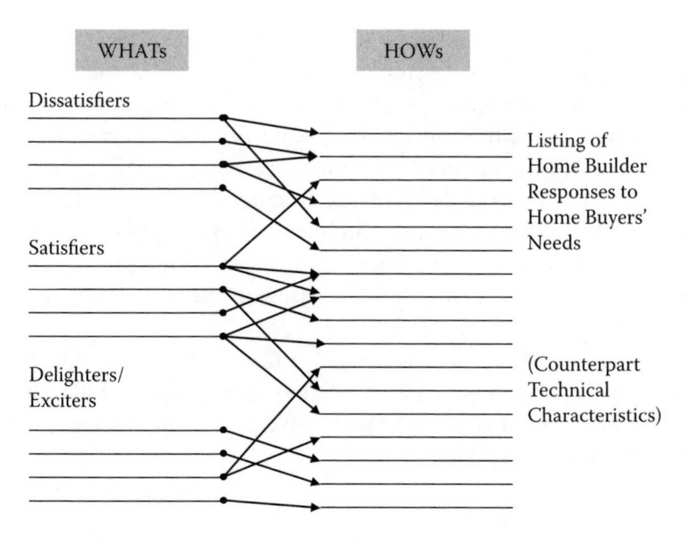

FIGURE 14.6 Organizational responses (HOWs).

require multiple HOWs such as increased square footage on all floors and especially clever layouts to ensure easy access to the closets from two or more points of egress.

It should also be noted that some HOWs are responsive to more than just one WHAT. An example HOW might include a low-voltage electrical system that is

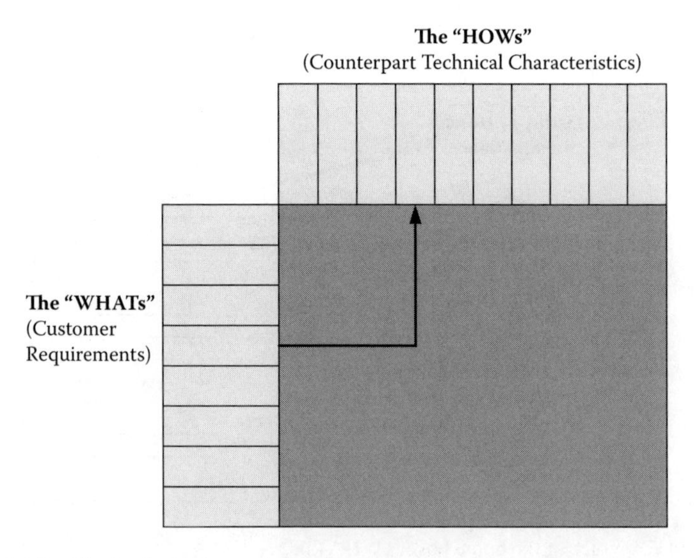

FIGURE 14.7 The basic QFD matrix: demonstrating the WHAT–HOW concept.

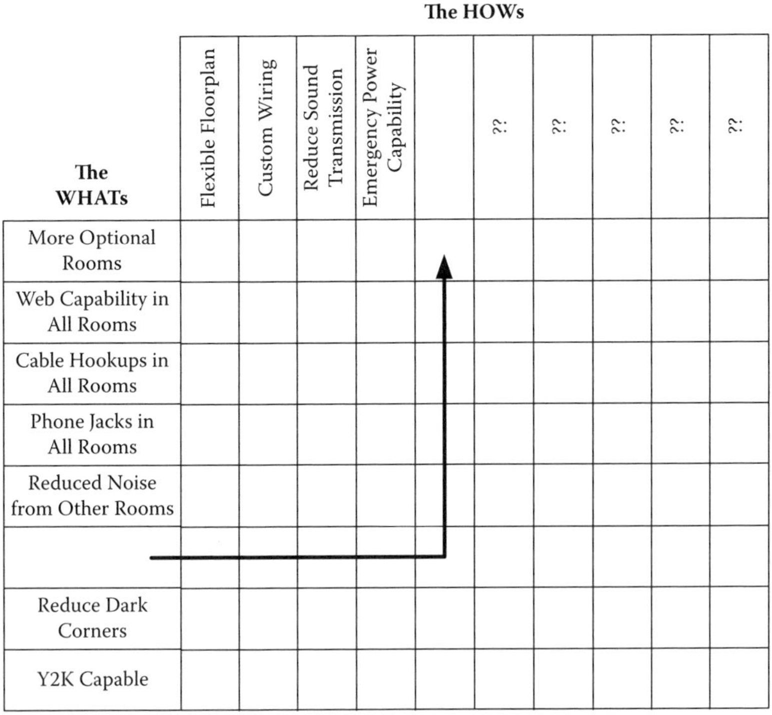

The HOWs

The WHATs	Flexible Floorplan	Custom Wiring	Reduce Sound Transmission	Emergency Power Capability		??	??	??	??	??
More Optional Rooms										
Web Capability in All Rooms										
Cable Hookups in All Rooms										
Phone Jacks in All Rooms										
Reduced Noise from Other Rooms										
Reduce Dark Corners										
Y2K Capable										

FIGURE 14.8 The house of quality QFD requirements matrix.

responsive to WHATs such as a home that is cable ready, has an internal speaker system, and has a built-in video-based security system.

The core interrelationship between the WHATs and the HOWs is reflected in Figure 14.7. This core interrelationship is taken a step further in Figure 14.8, which transfers the concept to the residential construction industry.

14.4 THE HOUSE OF QUALITY

Within the matrix of matrices, the initial QFD matrix—also known as the requirements matrix, the A-1 matrix (column A and row 1 in the matrix of matrices), and the product planning matrix—is commonly referred to as the HOQ. Figure 14.9 presents a blueprint of the HOQ. It is important to note that depending on the nature of each project, not every room in the HOQ is used in every project

Figure 14.10 demonstrates the waterfall relationship, that is, the cascading of information from one QFD matrix to another. This figure demonstrates that the output of each predecessor matrix becomes the input to each successor matrix.

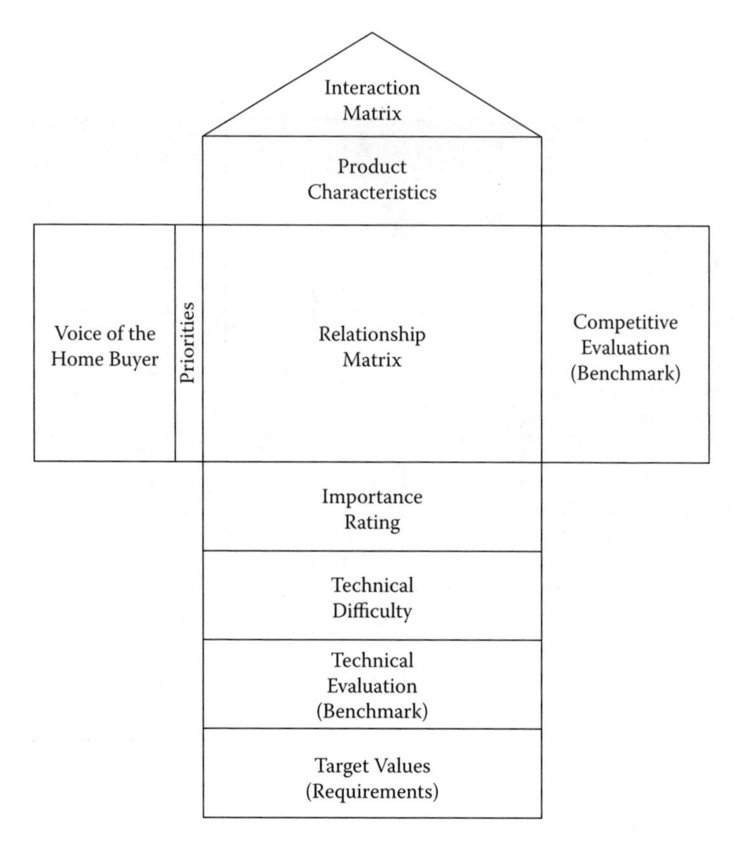

FIGURE 14.9 QFD HOQ basic blueprint.

Without completing each predecessor matrix, it is virtually impossible to develop its successor matrices.

A complete yet simple explanation of how the HOQ is constructed can be viewed on the Internet by going to the URL that follows. Dr. Robert Hunt of Macquarie University (Sydney, Australia) deserves due credit for the conceptualization and creation of this animated tutorial:

http://www.gsm.mq.edu.au/wps/wcm/connect/42b2da004a1193059276ff061c685d80/qfd-hoq-tutorial.swf?MOD=AJPERES&CACHEID=42b2da004a1193059276ff061c685d80

14.5 ADVANTAGES AND LIMITATIONS OF QFD

Figures 4.10 through 4.1 are provided so the reader can fully appreciate QFD's many strengths and can recognize its limitations. Figure 14.11 offers 10 advantages of using QFD, whereas Figure 14.12 discusses the nearly 20 benefits of QFD implementation.

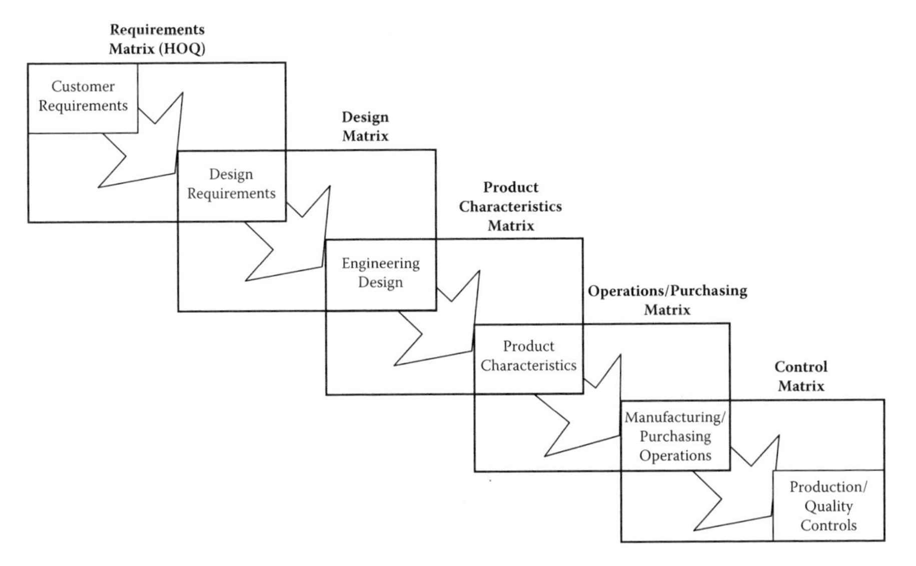

FIGURE 14.10 QFD waterfall relationships.

Home builders will be especially interested in Figure 14.13, which provides the results of applying QFD in residential construction. Finally, in the spirit of full disclosure, Figure 14.14 itemizes three potential limitations of QFD.

14.6 QFD SOFTWARE

As mentioned at the beginning of this chapter, QFD had its origins in Japan during the 1960s and 1970s and then found its way across the Pacific Ocean to the United

QFD:

- Is Systematic and Structured
- Assures Product Characteristics Equate to Customer Requirements
- Avoids Omissions Resulting from Oversight
- Avoids Under and Over Specification
- Identifies Specific Tools and Techniques That Produce the Greatest Payoffs
- Maps Company System to Customer Requirements
- Is as Sophisticated or Simplistic as You Care to Make It
- Identifies Important Characteristics That Must Be Controlled
- Results in Fewer Start-up Problems and Costs
- Produces a Detailed Command Media Documentation Trail

THUS, QFD ASSURES CUSTOMER SATISFACTION

FIGURE 14.11 Advantages of QFD.

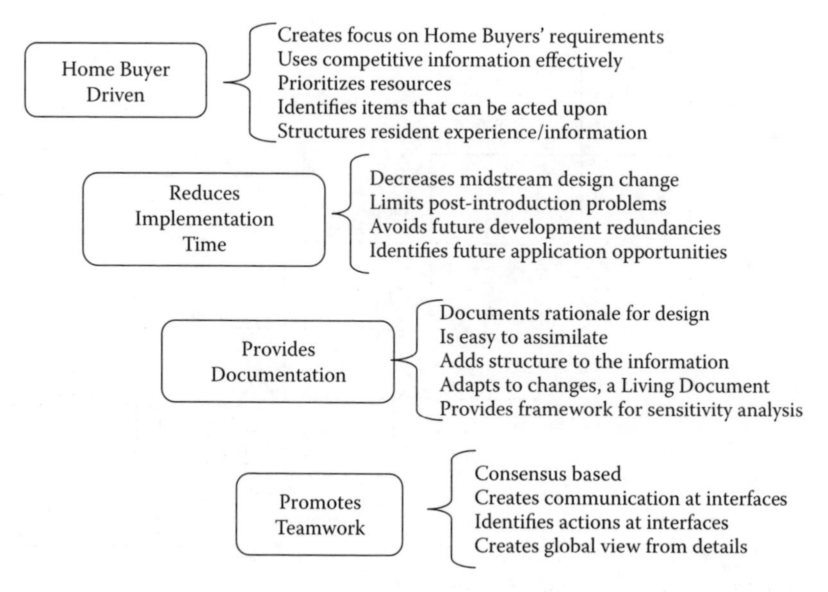

FIGURE 14.12 QFD benefits.

- Brings new (and carryover) designs to home buyer sooner than the competition with lower cost and improved quality
- Translates "voice of home buyer (customer)" into parameter designs for deployment throughout virtual organization to reduce product development time
- Identifies conflicting design requirements to be resolved with parameter design case studies using design of experiments (see Chapter 15)
- Provides system for design (of products or services) based on home buyer (customer) demands
- Involves all members of virtual organization
- Adds design to improvement & maintenance activities/responsibilities of all employees so home buyers receive builders' best output

FIGURE 14.13 Results of applying QFD in residential construction.

QFD:

- Requires team training of cadre
- Can be laborious if facilitator is inexperienced
- Requires use of computer software

FIGURE 14.14 Some limitations of QFD.

- QFD/Capture–International Technigroup, Inc. (ITI)
  (800) 783-9199
  www.qfdcapture.com

- QFD/Designer–Ideacore
  (248) 433-3380
  www.ideacore.com

- QFD/Pathway–"The QFD Handbook"
  (800) Call-Wiley
  www.amazon.com

- QFD/Builder–QFD OnLine
  www.qfdonline.com/qfd-builder-online-software/

FIGURE 14.15 QFD software.

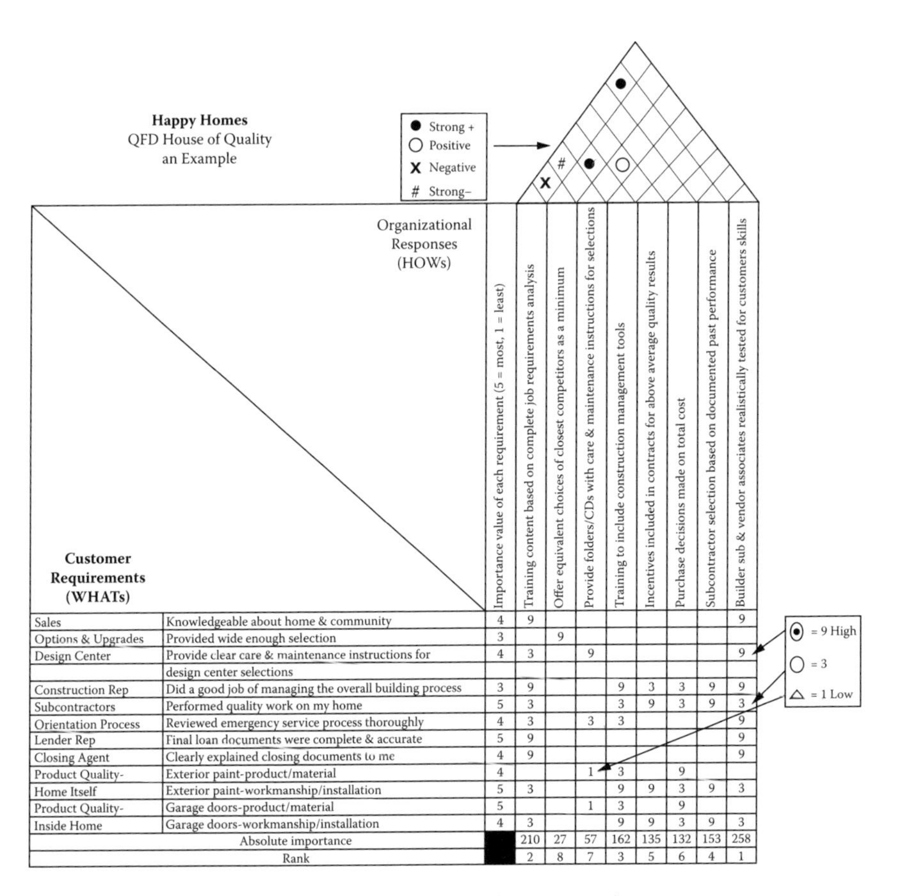

Customer Requirements (WHATs)		Importance value of each requirement (5 = most, 1 = least)	Training content based on complete job requirements analysis	Offer equivalent choices of closest competitors as a minimum	Provide folders/CDs with care & maintenance instructions for selections	Training to include construction management tools	Incentives included in contracts for above average quality results	Purchase decisions made on total cost	Subcontractor selection based on documented past performance	Builder sub & vendor associates realistically tested for customers skills
Sales	Knowledgeable about home & community	4	9							9
Options & Upgrades	Provided wide enough selection	3		9						
Design Center	Provide clear care & maintenance instructions for	4	3		9					9
	design center selections									
Construction Rep	Did a good job of managing the overall building process	3	9			9	3	3	9	9
Subcontractors	Performed quality work on my home	5	3			3	9	3	9	3
Orientation Process	Reviewed emergency service process thoroughly	4	3		3	3				9
Lender Rep	Final loan documents were complete & accurate	5	9							9
Closing Agent	Clearly explained closing documents to me	4	9							9
Product Quality-	Exterior paint-product/material	4			1	3		9		
Home Itself	Exterior paint-workmanship/installation	5	3			9	9	3	9	3
Product Quality-	Garage doors-product/material	5			1	3		9		
Inside Home	Garage doors-workmanship/installation	4	3			9	9	3	9	3
Absolute importance			210	27	57	162	135	132	153	258
Rank			2	8	7	3	5	6	4	1

FIGURE 14.16 Example of QFD HOQ in residential construction.

States in the 1980s. During these first two decades, all of the QFD matrices were created either by hand or using Excel software. Then, as QFD grew in popularity, several QFD-specific software packages were developed. As a result of this important assist, the use of QFD continued to expand to a broad variety of businesses and industries as well as across the world. Figure 14.15 lists the four most popular QFD software packages and how to locate them.

14.7 EXAMPLE

Figure 14.16 presents a QFD house of quality that might be used by a home builder.

15 Design of Experiments

15.1 APPROACHES TO EXPERIMENTAL DESIGN

Experimentation is the heart of continuous improvement. To improve a process or product, new things must be tried and evaluated. While this is a well-known concept, the actual implementation is not always appropriately implemented. The implementation of experimental design can be categorized into four approaches: haphazard, best guess, one factor at a time (OFAT), and design of experiments (DOE).

What is referred to as the haphazard approach is essentially making change after change outside the boundaries of specific controlled conditions. This approach is typified by identifying a problem or metric outside the acceptable boundaries and essentially making a decision to change one or more aspects to try and "fix" the problem. This may or may not lead to actual improvements. If the resulting metric remains outside acceptable boundaries, other changes should be made. What characterizes this approach is that the changes are based on intuition or some degree of experience, but are not systematic in nature and almost always seems random. This approach is not recommended as it has a low probability of sustained success and is very inefficient and potentially quite costly.

The best-guess method is only slightly better than the haphazard approach. It involves identifying factors of interest and testing a baseline scenario. Based on the results of the baseline scenario, adjustments are made to the factors based on knowledge and experience. While it is more systematic than the haphazard approach, it is still a very inefficient long-term approach. In any individual project it might be very successful, but done over a number of projects the best-guess approach results in far more tests and expenses than is really necessary.

The one-factor-at-a-time approach is widely practiced and even thought to be scientifically robust in nature. While it may be an improvement over the two previously illustrated approaches, several significant drawbacks are associated with the OFAT approach. We suggest that it is not as scientifically robust as most people think. Using this approach, a problem is evaluated by identifying the key/critical factors and manipulating each of them, one at a time, to determine the outcome. While this seems like a logical thing to do, it does not take into account the interactions that occur between the factors. An interaction takes place when the effect of one factor set at a particular level depends on the state or level of another factor. In other words, there may be situations where the effect of factor A, having been set at level 1, depends on the setting of factor B. These situations are not uncommon and should be estimated and accounted for in the appropriate situations. The problem with the OFAT approach is that it will never account for these situations because it is not able to do so.

Using the DOE approach, considered to be most effective and efficient, the effect of each factor as well as the effects of the potential interactions of the factors can be estimated effectively and efficiently using the fewest number of test situations. Of course the greater the number of factors and factor interactions, the greater the expense to implement this approach.

15.2 OBJECTIVES OF DOE

DOE is one of the most powerful and flexible tools available for continuous improvement because of its efficiency and flexibility. It can be used to improve production processes as well as the products themselves. Home builders should be aware that DOE can be used for, but is not limited to, the following objectives:

1. Determine which of many input factors have a significant impact on a specified output variable.
2. Determine the effect that specific input factors have on a specified output variable.
3. For each influential input factor, determine its optimal level/setting so the output variable is almost always near the desired target value.
4. Determine what level/setting should be used for each influential factor to minimize the variability in the specified output variable.
5. Determine what level/setting should be used for each influential factor so that the undesirable effects of the uncontrollable or noise influences are minimized.

The objective of the research will be the most significant factor in determining which type of design should be used. The experimental design can range from simple to complex while maintaining unbiased results. Given the complexity of the subject, this chapter is designed to provide an overview of the concepts and applications available rather than a detailed review of the mechanics and calculations.

15.3 FACTORIAL DESIGNS

The workhorse design in DOE is called the factorial design. The factorial design has a number of properties that make it both incredibly powerful in its ability to detect effects or outcomes and the building block for other experimental designs with different objectives. The fundamental factorial design is the 2×2 (read "two by two") factorial. The notation of the design indicates that there are two levels or settings for two unique factors being manipulated. For example, if the objective was to improve control of water temperature in the home, two potential input factors could be material and placement of water pipes. In a 2×2 scenario, the material factor could have the two levels of copper and cross-linked polyethylene (PEX; the tubing widely used for plumbing and radiant heating applications), while the placement could be under the foundation or through the attic. In this situation the experimenter would be manipulating two factors (material and placement), and each factor would have two levels (copper versus PEX and under foundation versus through attic). The experimental design would

		Material	
		Copper	PEX
Placement	Under the Foundation	W	X
	Through the Attic	Y	Z

FIGURE 15.1 Experimental design (four-cell box).

result in four possible combinations as shown in Figure 15.1. Each cell contains a letter that represents the quantitative value of whatever performance metric is being used to assess the material–placement combination. For example, *W* represents the result of an experimental run when copper tubing was placed under the foundation.

By testing all four conditions in a randomized order, the factorial design is a powerful tool in detecting both the direct or main effects of materials and placement, as well as any interaction that may exist between them. Depending on the output variable being measured, it may be necessary to replicate the design with multiple observations for each combination. In other instances it is possible to estimate effects with just one or two observations per cell or combination.

Factorial designs are not limited to just two factors or even two variables per factor. A 2×3 factorial design would indicate that there are two levels per factor and three unique factors being manipulated. The trade-off for increases in the number of factors or levels being tested is that it requires more observations to test. In the 2×2 design there are four different combinations to test; in the 2×3 design the combinations increase exponentially, resulting in eight combinations to test. Likewise, a 2×4 design would result in 16 combinations to test. Figure 15.2 illustrates a 2×3 factorial design.

Factorial designs, while powerful, are not necessarily ideal for every objective. In industrial settings, there may be a need to test three, four, or five variables and sometimes even more. The number of observations needed in these situations increases dramatically. All of these observations have a related cost in material, labor, and time associated with them that can rapidly escalate the cost of the experiment. Using one of the unique properties of factorial designs called projection led to the development of the fractional factorial design best suited to address this issue.

15.4 FRACTIONAL FACTORIAL DESIGNS

While a factorial design investigates every possible combination of factors individually, the fractional factorial design intentionally trades the estimation of higher order (more complicated) interaction effects to focus on the main effects and lower level (less complicated) interactions of the experiment. In doing this, the experiment can test the same number of factors with only a fraction (e.g., one half, one quarter, one eighth, one sixteenth) of the observations or combinations to test. This concept is demonstrated in Figure 15.3.

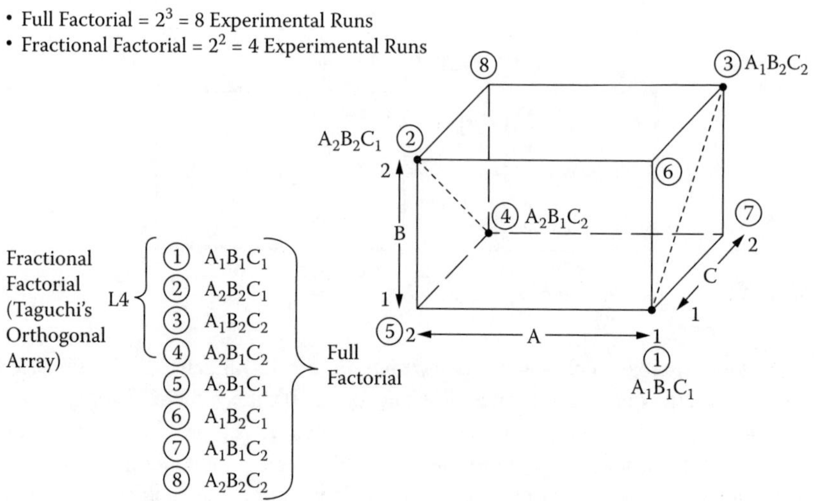

FIGURE 15.2 Example of three factors at two levels.

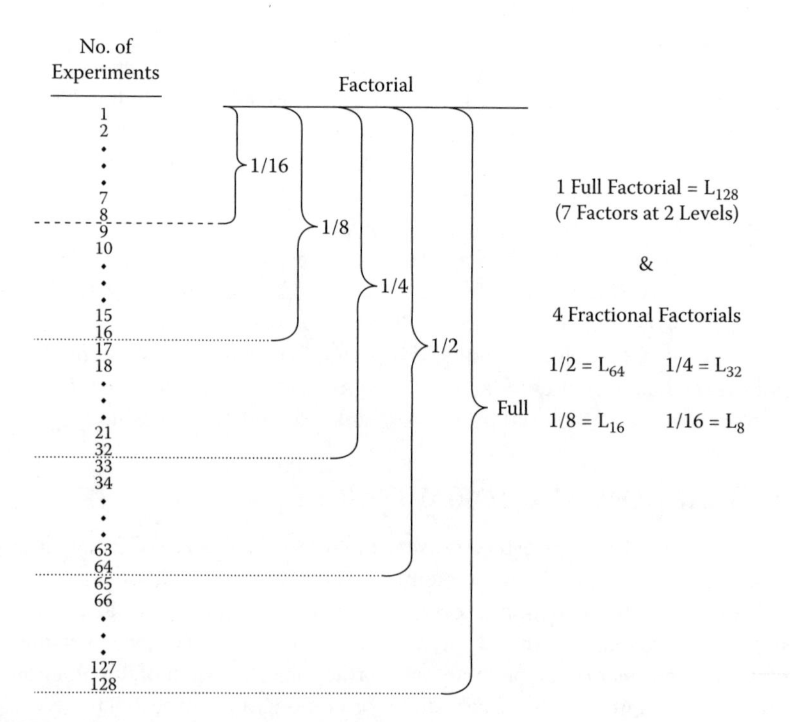

FIGURE 15.3 Full versus fractional factorial designs.

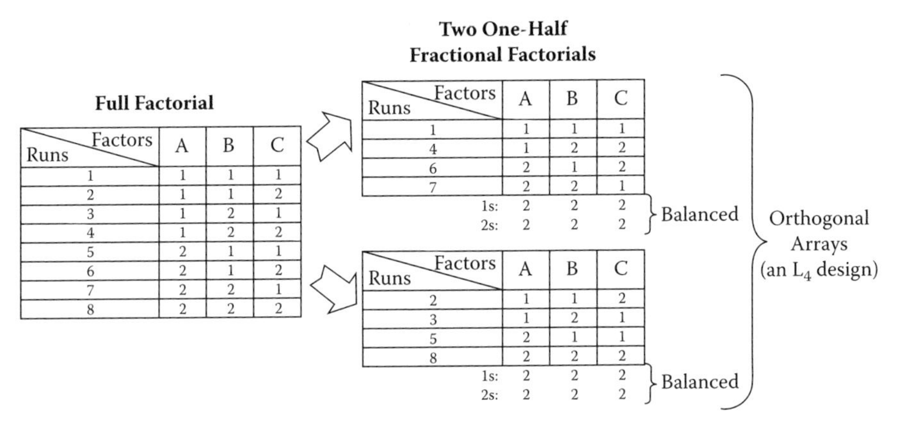

FIGURE 15.4 Full factorial design versus fractional factorial design.

The number of observations eliminated from the full factorial design depends on how many effects the experimenter is willing to live without. This trade-off is referred to as the design resolution. To illustrate this, let's use a 2×3 factorial design, which has three different factors being manipulated at two different levels or values. This experimental design is pictured on the left side of Figure 15.4. As stated previously, this produces eight total combinations (experimental runs) to test in the full factorial design. Using a one-half fractional factorial design, the main effects of all three factors could be estimated using four combinations instead of the full eight. What's being given up are the interaction effects between the factors and the corresponding knowledge that could result. If the interaction effects are of little or no interest given the situation, efficiency, time, and savings gained using the fractional design could be quite valuable.

15.5 SCREENING DESIGNS

The efficiency gained from the fractional factorial design is especially true when dealing with much larger designs. If the situation called for testing 10 different variables, each with two levels, the full factorial design would require 1,024 unique observations to estimate all the parameters. However, using a 1/32 fractional design the same experiment could be run in 32 observations and still produce reliable estimates for the main effects and the basic (lower-level) interactions.

These larger designs are generally used as screening designs. These are designs that were developed to accommodate a large number of factors that may or may not actually be important to the output. The screening design can be used in problem identification by helping weed out factors that have no impact on the problem process or product.

For illustrative purposes, let's use stucco cracking as an example. Stucco is a plaster-type coating commonly used on the exterior of homes in southern and southwestern United States. While more suited to the rigors of the local environment than wood, stucco isn't perfect. It does crack over time, but the extent of cracking can be

Data Collection Sheet: Stucco Experiment

BACKGROUND

Community ____________ Lot ________ Plan ________ Elevation ________

Quality Assurance Representative ____________________

HOME DETAILS

Typical Exterior Stud Size ____________

Dates: Final Plaster (Texture) ____________

Paint ____________

Inspection (QA) ____________

FACTORS

Please circle the correct factor level for each controllable and uncontrollable factor.

Controllable Factors	**Factor Levels** 1	2
Foundation System	Semi-monolithic	Post Tension
Wall Framing	Stick or Component	Panel
Time Between Stucco Coats	1 or 2 days	3 or more days
Painting	Spray	Backroll
Ceiling Profile	Flat	Vaulted (any ceiling)
Foam Joints	Taped	Caulked
Lath Quality	Excellent	Acceptable

Uncontrollable Factors	**Factor Levels** 1	2
No. of Stories	One	Two
Season	May–September	October–April
Wind Speed	High	Low

CRACKS

Please report the extent of stucco cracking observed from 5 feet away.

Crack Length	**Crack Thickness**	
________ feet	Thinner Than a Dime	Equal to or Thicker Than a Dime

Please circle the compass orientation of the wall with the most cracks.

Compass Orientation

North East West South

FIGURE 15.5 Stucco screening experiment.

controlled. The question on the table is, "What factors contribute to stucco's propensity to crack?" Is it water content, heat, application methods, mixture, dry materials, or something else? It's easy to see how the list of potential factors can grow very quickly. Screening designs allow the experimenter to identify the important factors, so the irrelevant ones do not waste time and resources. Figure 15.5 offers a look at such an experiment.

Plackett–Burman designs are for situations that require screening even larger numbers of factors. These designs all require runs to be multiples of four and can evaluate $n - 1$ factors (where n is the number of runs or observations). For example, using a Plackett–Burman design with 12 runs/observations, the experimenter can test the main effects of up to 11 different factors. With 20 runs or observations, the design can estimate the effects of up to 19 different factors. Again, these designs hinge on the assumption that the interaction between factors is negligible or of little interest. It's worth noting that many statistical software packages (such as those discussed in Chapter 12) include these designs to aid in the creation of experimental designs and analysis of the resulting data. However, it is also worth noting that it is best to have someone with sufficient DOE training and experience involved in leading this effort to minimize the time, cost, and effort associated with such experiments.

15.6 ROBUST DESIGN/TAGUCHI METHODS

Another common objective is to develop a product or process that minimizes the effects of uncontrollable or noise influences that degrade product and process quality or cause any of a number of problems. This type of experimental design is called a robust design or Taguchi method. To continue the stucco example previously illustrated, some of the factors may be outside of your control, such as daytime temperatures, humidity, or wind conditions. The objective is to find the ideal combination of controllable factors that minimizes the amount of cracking after a specified period of time. When this combination of controllable factors is found, it is referred to as being robust (or insensitive) to the effects of the uncontrollable or noise factors. In this scenario the experiment could be designed as in Figure 15.6.

Dr. Genichi Taguchi created a variety of robust design templates that were structured using orthogonal arrays; these include two, three, four, five, and even more levels for the factors being tested. Figure 15.7 demonstrates three examples of two-level

Controllable Factors	Factor Levels	
Foundation system	Semi-monolithic	Post tension
Wall framing	Component	Panel
Time between stucco coats	1–2 days	3 or more days
Painting method	Spray	Backroll
Ceiling profile	Flat	Vaulted (any ceiling)
Foam joints	Taped	Caulked
Lath quality	Excellent	Acceptable

Uncontrollable Factors	Factor Levels	
Number of stories	1-story	2-story
Season	May–September	October–April
Wind speed	High (15 or more mph)	Low (5 or less mph)

FIGURE 15.6 Robust DOE to reduce stucco cracking.

$L_4\ (2^3)$

Run No. \ Column	1	2	3
1	1	1	1
2	1	2	2
3	2	1	2
4	2	2	1
GROUP	1	2	2

$L_8\ (2^7)$

Run No. \ Column	1	2	3	4	5	6	7
1	1	1	1	1	1	1	1
2	1	1	1	2	2	2	2
3	1	2	2	1	1	2	2
4	1	2	2	2	2	1	1
5	2	1	2	1	2	1	2
6	2	1	2	2	1	2	1
7	2	2	1	1	2	2	1
8	2	2	1	2	1	1	2
GROUP	1	2	2	3	3	3	3

$L_{16}\ (2^{15})$

Run No. \ Column	1	2	3	4	5	6	7	8	9	10	11	12	13	14	15
1	1	1	1	1	1	1	1	1	1	1	1	1	1	1	1
2	1	1	1	1	1	1	1	2	2	2	2	2	2	2	2
3	1	1	1	2	2	2	2	1	1	1	1	2	2	2	2
4	1	1	1	2	2	2	2	2	2	2	2	1	1	1	1
5	1	2	2	1	1	2	2	1	1	2	2	1	1	2	2
6	1	2	2	1	1	2	2	2	2	1	1	2	2	1	1
7	1	2	2	2	2	1	1	1	1	2	2	2	2	1	1
8	1	2	2	2	2	1	1	2	2	1	1	1	1	2	2
9	2	1	2	1	2	1	2	1	2	1	2	1	2	1	2
10	2	1	2	1	2	1	2	2	1	2	1	2	1	2	1
11	2	1	2	2	1	2	1	1	2	1	2	2	1	2	1
12	2	1	2	2	1	2	1	2	1	2	1	1	2	1	2
13	2	2	1	1	2	2	1	1	2	2	1	1	2	2	1
14	2	2	1	1	2	2	1	2	1	1	2	2	1	1	2
15	2	2	1	2	1	1	2	1	2	2	1	2	1	1	2
16	2	2	1	2	1	1	2	2	1	1	2	1	2	2	1
GROUP	1	2	2	3	3	3	3	4	4	4	4	4	4	4	4

FIGURE 15.7 Taguchi's two-level orthogonal arrays.

orthogonal arrays, while Figure 15.8 contains Dr. Taguchi's triangular table used in the creation of these and other two-level orthogonal arrays.

15.7 OPTIMIZATION DESIGNS

The final type of design covered in this chapter focuses on determining the optimal settings of each factor to maximize the output variable. These designs have proven to be quite useful in both product and process development. While these may seem to be completely unrelated to residential construction, in reality the applications are not that different. Take concrete as an example. Concrete has a number of properties such as hardness or slump that are critical to the construction process. Concrete is a mixture of a number of ingredients mixed together in a certain ratio. The mixture and ingredients

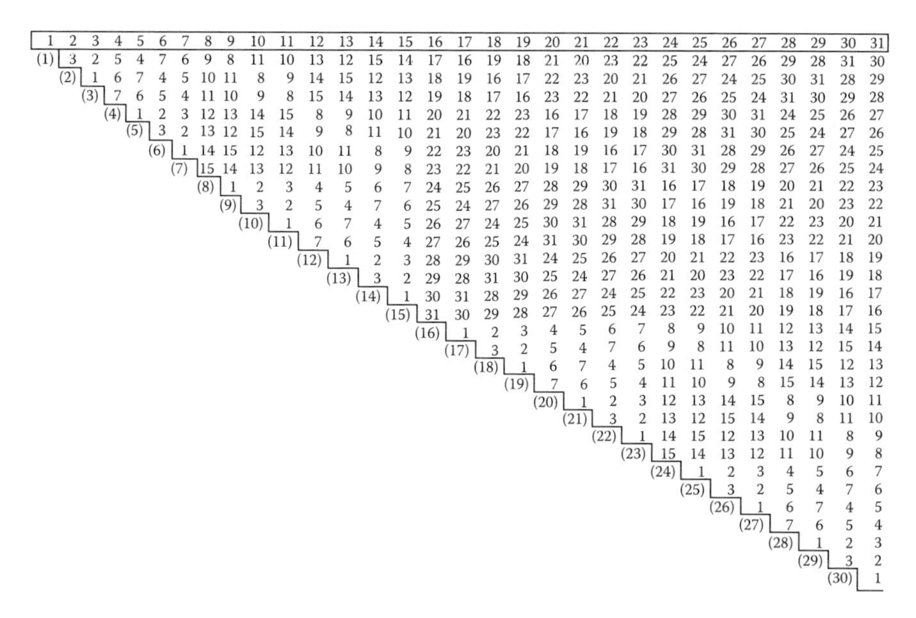

	1	2	3	4	5	6	7	8	9	10	11	12	13	14	15	16	17	18	19	20	21	22	23	24	25	26	27	28	29	30	31
(1)		3	2	5	4	7	6	9	8	11	10	13	12	15	14	17	16	19	18	21	20	23	22	25	24	27	26	29	28	31	30
(2)			1	6	7	4	5	10	11	8	9	14	15	12	13	18	19	16	17	22	23	20	21	26	27	24	25	30	31	28	29
(3)				7	6	5	4	11	10	9	8	15	14	13	12	19	18	17	16	23	22	21	20	27	26	25	24	31	30	29	28
(4)					1	2	3	12	13	14	15	8	9	10	11	20	21	22	23	16	17	18	19	28	29	30	31	24	25	26	27
(5)						3	2	13	12	15	14	9	8	11	10	21	20	23	22	17	16	19	18	29	28	31	30	25	24	27	26
(6)							1	14	15	12	13	10	11	8	9	22	23	20	21	18	19	16	17	30	31	28	29	26	27	24	25
(7)								15	14	13	12	11	10	9	8	23	22	21	20	19	18	17	16	31	30	29	28	27	26	25	24
(8)									1	2	3	4	5	6	7	24	25	26	27	28	29	30	31	16	17	18	19	20	21	22	23
(9)										3	2	5	4	7	6	25	24	27	26	29	28	31	30	17	16	19	18	21	20	23	22
(10)											1	6	7	4	5	26	27	24	25	30	31	28	29	18	19	16	17	22	23	20	21
(11)												7	6	5	4	27	26	25	24	31	30	29	28	19	18	17	16	23	22	21	20
(12)													1	2	3	28	29	30	31	24	25	26	27	20	21	22	23	16	17	18	19
(13)														3	2	29	28	31	30	25	24	27	26	21	20	23	22	17	16	19	18
(14)															1	30	31	28	29	26	27	24	25	22	23	20	21	18	19	16	17
(15)																31	30	29	28	27	26	25	24	23	22	21	20	19	18	17	16
(16)																	1	2	3	4	5	6	7	8	9	10	11	12	13	14	15
(17)																		3	2	5	4	7	6	9	8	11	10	13	12	15	14
(18)																			1	6	7	4	5	10	11	8	9	14	15	12	13
(19)																				7	6	5	4	11	10	9	8	15	14	13	12
(20)																					1	2	3	12	13	14	15	8	9	10	11
(21)																						3	2	13	12	15	14	9	8	11	10
(22)																							1	14	15	12	13	10	11	8	9
(23)																								15	14	13	12	11	10	9	8
(24)																									1	2	3	4	5	6	7
(25)																										3	2	5	4	7	6
(26)																											1	6	7	4	5
(27)																												7	6	5	4
(28)																													1	2	3
(29)																														3	2
(30)																															1

FIGURE 15.8 Taguchi's triangular table for two-level orthogonal arrays.

used, such as water content, aggregate size, fly ash substitutes, and even curing conditions, determine the characteristics of the concrete. If the objective is to optimize a particular characteristic such as hardness of the concrete after 10 days, a variation of the factorial design could be used. This variation is called a central composite design.

The difference between a factorial design and a central composite design is that it uses center points and axial runs. These additional points, as shown in Figure 15.9,

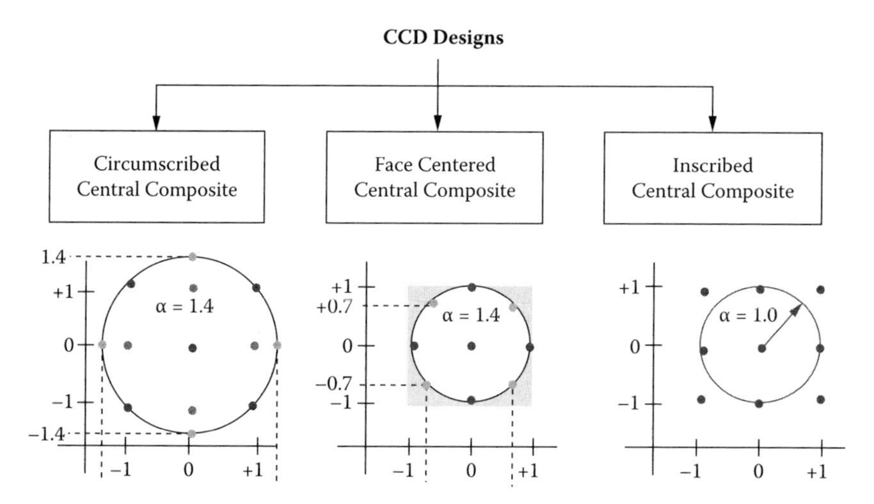

FIGURE 15.9 Information function for two-factor rotatable central composite design.

allow the experimenter to test for and estimate curved relationships between variables and to identify the values necessary for each factor to produce the maximum effect.

It is important to note that there may be more than one optimal solution when the experiment is run. After identifying the list of combinations, the next step would be to identify which of the combinations is the best fit for the existing project.

15.8 PLANNING AND RUNNING DOE

As with any type of research, planning is the most important aspect. No type of data analysis can make up for a poorly planned experiment. With that in mind, certain aspects of the experiment should be well thought out prior to proceeding with the actual test.

First, the experiment should have a clear and well-defined objective. The objective will help determine the appropriate research design to be used. This definition extends both to the problem to be solved as well as to the factors to be tested. The defined objective should also include what will be used in the experiment and how the results will be used or will impact the business.

The experiment should also be planned so that outside influences (noise variables) are minimized. This means that if the intent is to estimate the effect of certain factors on a process or product, the experiment should be run in conditions where all of the other factors are held relatively constant. In reality, it is virtually impossible to control every outside factor to exactly match the conditions for each test, but through other techniques such as randomization of test order and replication of the observations these factors can be accounted for in the analysis so that the results are not biased.

15.9 TYPES OF EXPERIMENTS

As stated earlier in this chapter, two main types of experiments are found in the existing literature: full factorial experiments and fractional factorial experiments. The pros and cons of these experiments have already been discussed and will not be covered again. However, other types of DOEs are frequently mentioned in other writings.

Before discussing the details of these other types, let's look at Figure 15.10. We see a Venn diagram with three overlapping circles. Each circle represents a specific school of thought or approach to designed experiments: (1) classical methods (one thinks of Drs. George Box or Douglas Montgomery), (2) Taguchi methods (referring to Dr. Taguchi), and (3) statistical engineering (established and taught by Mr. Dorian Shainin). We also see that all three approaches share a common focus, that is, the factorial principle referred to earlier in this chapter. The figure further demonstrates that each pairing of approaches shares a common focus or orientation, one approach with another. Finally, it is clear that each individual approach possesses its own unique focus or orientation.

Another way to differentiate among classical DOE, Taguchi methods (robust design), and Shainin methods (statistical engineering) is to use a comparative analysis in the form of a matrix such as that shown in Figure 15.11.

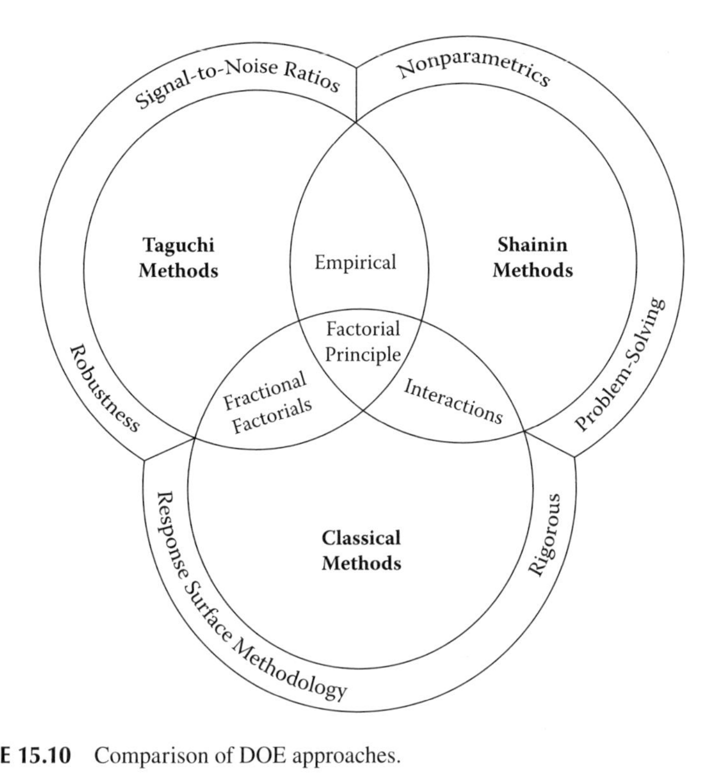

FIGURE 15.10 Comparison of DOE approaches.

The type of nonclassical experiment most often discussed is the Taguchi method, also called robust design and occasionally quality engineering. Taguchi experiments are fractional factorial experiments. In that regard, Taguchi's experimental structures are not significantly different from their classical design counterparts as is his presentation of experimental arrays and his approach to the analysis of results. Some practicing statisticians do not promote Dr. Taguchi's experimental arrays due to opinions that other experimental approaches are superior. Despite this, many knowledgeable DOE professionals have noted that practicing engineers seem to grasp experimental methods as presented by Dr. Taguchi more readily than methods advocated by classical statisticians (Drs. Box and Montgomery) and quality engineers (Mr. Shainin). It may be that Dr. Taguchi's use of graphical analysis contributes to this phenomenon. Although analysis of variance (ANOVA) and regression are strongly grounded in statistics and are very powerful, telling an engineer what factors and interactions are important is less powerful than showing him or her the direction of effects through the use of graphical analysis.

Despite the relatively small controversy regarding Taguchi methods, Dr. Taguchi's contributions to DOE thinking remain intact. This influence runs from the promotion of his experimental tools (such as the signal-to-noise ratio, loss function, and orthogonal array) to perhaps more importantly, his promotion of the use of designed experiments to reduce the influence of product and process variation

How → / What ↓	Classical Design of Experiments		Taguchi Robust Design	Shainin Statistical Engineering
	Full Factorial	Fractional Factorial		
New Design		○	●	
Many Factors		○	●	△
Few Factors	●		○	●
Known Interactions	●	○	○	●
Few Interactions	○		●	○
Inexpensive Test Run	●		●	●
Costly Test Run	△	○	* ●	
VRP		○	●	●
Process Centric			○	●
Improvement	○		●	●
High Reliability Required	●	△	○	●
Failure Analysis		△	○	●

Best ●
Good ○
Fair △

VRP = Variability Reduction Process

* Taguchi Usually Followed by Confirmation Run

FIGURE 15.11 DOE applications table.

as well as uncontrollable factors. Dr. Taguchi described uncontrollable factors, often called noise factors, as elements in a process that are too costly or too difficult or perhaps even impossible to control. A classic example of an uncontrollable factor is copier paper. Despite instructions and specifications to the contrary, a photocopier customer will use whatever paper is available, especially when a deadline is nearing. If the wrong paper is used and a jam is created, the photocopier service personnel would be correct to point out the error that the user was not following specifications. Unfortunately, the customer will still be dissatisfied. Dr. Taguchi recommends making the copier's internal processes more robust (insensitive) against paper thickness variation, one of the uncontrollable factors.

15.10 BEFORE THE STATISTICIAN ARRIVES

Most residential construction organizations that have not yet instituted the use of Six Sigma have few, if any, persons with much knowledge of applied statistics. To support this type of organization, it is suggested that process improvement teams (PITs) make use of the following process to help them define, measure, analyze, improve, and control (DMAIC). Figure 15.12 identifies the various steps PITs should take to maximize the usefulness of their statistical mentors.

1. Form a process improvement team (PIT) of subject matter experts (SMEs) from all company functions that are related to the problem. These are referred to as stakeholders.
2. Describe the problem in under 100 words.
3. State the objective of the experiment:
   a. Define at least one quality/performance characteristic to be measured/counted. If there is more than one, prioritize them from most to least important.
   b. With respect to each characteristic:
      (1) Identify whether bigger is better, smaller is better, or nominal is best as well as how data will be bundled, e.g., DPU, DPDU, yield, fraction defective, C_p, C_{pk}, or Six Sigma.
      (2) Determine the method to be used to collect the desired data, i.e., how to measure/count the characteristics.
4. List and then prioritize (as objectively as is possible) those controllable input variables that are believed to be significant with respect to the output quality/performance characteristic. (Refer to 3a above) NOTE: The PIT should be advised that the more variables to be investigated, the larger the number of units that will be needed for their experiment.
5. List the levels of interest for each controllable input variable. (Refer to 4 above.) At the outset, two levels are preferable, but no more than three for a screening experiment. For example, if stud angle from the vertical is an input variable of interest, the levels might be: Level 1 = 0 degrees, Level 2 = 2 degrees, and Level 3 = 4 degrees. One level should be the actual value presently in use.
6. List and then prioritize those uncontrollable input variables that are believed to be significant with respect to the output quality/performance characteristic, but which are beyond worker capability to regulate.
7. In order to desensitize the output relative to the input, list two or three levels of interest for each uncontrollable input variable.
8. The consulting statistician will then meet with the DOE PIT to discuss its written responses to items 1 to 7.
9. The consulting statistician will design/select a recommended experimental design for use by the PIT. Interaction between the PIT and the statistician is encouraged to facilitate fine tuning of the experiment as well as final agreement.
10. The DOE PIT should then conduct the experiment and provide the written results to the consulting statistician who will perform an analysis of variance (ANOVA) and then render a report to the PIT.
11. Based upon the report received from the statistician, the DOE PIT will draw conclusions and produce a written report of their experiment for review and approval by their senior management.

FIGURE 15.12 What to do before the statistician arrives.

15.11 ORTHOGONAL ARRAY

Orthogonal array is a technique that is important for organizations using or planning to use robust design or DOE, as discussed earlier in this chapter.

In the course of designing a simple experiment, suppose a team has three factors that need to be studied at two levels each. Using the common practice of factorial design, the number of experimental runs needed for a full factorial design is equal to

The Full Factorial Design Is an Array of Values Arranged as a Matrix:

Runs \ Factors	A	B	C
1	1	1	1
2	1	1	2
3	1	2	1
4	1	2	2
5	2	1	1
6	2	1	2
7	2	2	1
8	2	2	2

FIGURE 15.13 Full factorial design without interaction columns.

the number of factor levels (in this case = 2) raised to the power of the number of factors (in this case = 3). Two raised to the third power, or 2^3, equals eight experimental runs. The layout of the eight runs is as follows.

The 1s and 2s in the factor columns are factor levels. Thus, in Run 1, Factors A, B, and C are all at level 1; in Run 4, Factor A is at level 1 but Factors B and C are at level 2; and in Run 7, Factor C is at level 1 but Factors A and B are at level 2. This arrangement of factors and factor levels is portrayed in Figure 15.13.

A full factorial design is used when an experimenter needs to study both main effects (the results of just the factors) and interaction effects (the results of combining two or more factors). When an experimenter needs to study only the main effects or a combination of the main effects and only some of the interactions, then a full factorial design is not required and a fractional factorial design is selected.

In the preceding three factor-two level example, any four runs would constitute a one-half fractional factorial (e.g., Runs 1, 2, 3, and 4; or 3, 4, 5, and 6; or 1, 4, 6, and 7), as in Figure 15.4. These are all one-half fractional factorials because they have half as many runs as the full factorial from which they were drawn. In the third set of runs, (i.e., 1, 4, 6, and 7), we have a fractional factorial with some special properties; this is referred to as an orthogonal array. When this set is studied, the following properties are evident:

Special Property 1: An equal number of 1s and 2s must exist in each column. This balance is one of the three properties that must be present for a fractional factorial to qualify as an orthogonal array.

Special Property 2: Looking at any factor (e.g., Factor A), for those runs when the levels are the same (e.g., Runs 1 and 4 when Factor A is at level 1), then the frequency of occurrence of the levels of the other factors (in this case, levels 1 and 2 for Factors B and C) must be equal. Note there is a single 1 and a single 2 (an equal frequency) in the remaining columns (Factors B and C).

Special Property 3: To qualify as an orthogonal array, a fractional factorial must be the smallest one (i.e., the one with the least number of runs) for a given number of factors and factor levels and it must possess the first two special properties.

Full Factorial Design

- A full factorial design includes all factors & factor interactions (2-way interactions between 2 factors, 3-way interactions between 3 factors, etc.)
- A complete full factorial for 3 factor-2 level factorial design includes all factor interactions & is known as an L_8 design

Factors & Factor – Interactions

Runs \ Factors	A	B	A×B	C	A×C	B×C	A×B×C
1	1	1	1	1	1	1	1
2	1	1	1	2	2	2	2
3	1	2	2	1	1	2	2
4	1	2	2	2	2	1	1
5	2	1	2	1	2	1	2
6	2	1	2	2	1	2	1
7	2	2	1	1	2	2	1
8	2	2	1	2	1	1	2

Factor & Factor – Interaction Levels

FIGURE 15.14 Full factorial design with interaction columns.

These orthogonal arrays are important for use by experimenters because they permit studies to be conducted without having to pay for more runs than are actually necessary. Looking at Figure 15.14, if there were no need to study one or more of the three two-way interactions (A × B, A × C, and B × C) or the one three-way interaction (A × B × C), then it doesn't make sense to use a full factorial when a one-half fractional factorial can provide all the needed information.

In this case, the orthogonal array design costs half as much and takes half as long to conduct because only half the runs are necessary. In cases where an experimenter is studying more factors and factor levels than in this example, it is quite common to use an orthogonal array that is 1/16 of its corresponding full factorial. Clearly a major savings in cost and time can be realized.

Last, but certainly not least, there is still another important advantage in using an orthogonal array. When a DOE is performed using an orthogonal array, the results of the experiment can identify a combination of factors and factor levels that yield the best results (in terms of whatever performance measures or metrics are used), even if that combination was not one contained in the original orthogonal array. Yes, that's right; one of the attributes of an orthogonal array is that even when the optimal combination of factors and factor levels was not one of the experimental test runs, the subsequent analysis of the data resulting from the experiment will reveal it to be the best selection.

For example, in the foregoing case the DOE might identify Factors A, B, and C all at level 2 as the optimal design combination even though that combination was not part of the experiment. Recall that it was a part of the original full factorial but not a part of the subsequent fractional factorial (orthogonal array).

15.12 GLOSSARY OF DOE TERMS

Because DOE is such a specialized improvement technique, a glossary of associated terms is provided in Table 15.1.

15.13 EXAMPLE

Builder A has identified the need to reduce construction cycle time for its medium-sized homes. The PIT has identified four potential ways to improve cycle time. The first potential solution would be to reduce the workload or number of homes each field superintendent is carrying. This would require hiring additional people

TABLE 15.1
Glossary of DOE Terms

Term	Definition
Confounding	When a design is used that does not explore all the factor-level setting combinations, some interactions may be mixed with each other or with experimental factors such that the analysis cannot tell which factor contributes to or influences the magnitude of the response effect. When responses from interactions/factors are mixed, they are said to be *confounded.*
DOE	Design of experiments; also known as industrial experiments, experimental design, and design of industrial experiments.
Factor	A process setting or input to a process. For example, the temperature setting on an oven is a factor as well as the type of raw material used.
Factor-level settings	The combination of factors and their settings for one or more runs of the experiment. For example, consider an experiment in three factors, each with two levels (H and L = high and low). The possible factor-level settings include H-H-H and H-L-L.
Factor space	The hypothetical space determined by the extremes of all the factors considered in the experiment. If there are k factors in the experiment, the factor space is k-dimensional.
Interaction	Factors are said to have an interaction when changes in one factor cause an increased or reduced response to changes in another factor or factors.
Randomization	After an experiment is planned, the order of the runs is randomized. This reduces the effect of uncontrolled changes in the environment such as tool wear, chemical depletion, or warm-up.
Replication	When each factor-level setting combination is run more than one time, the experiment is *replicated.* Each run beyond the first one for a factor-level setting combination is a *replicate*.
Response	The result to be measured and improved by the experiment. In most experiments there is one response, but it is certainly possible to be concerned about more than one response.
Statistically significant	A factor or interaction is said to be statistically significant if its contribution to the variance of the experiment appears to be larger than would be expected from the normal variance of the process.

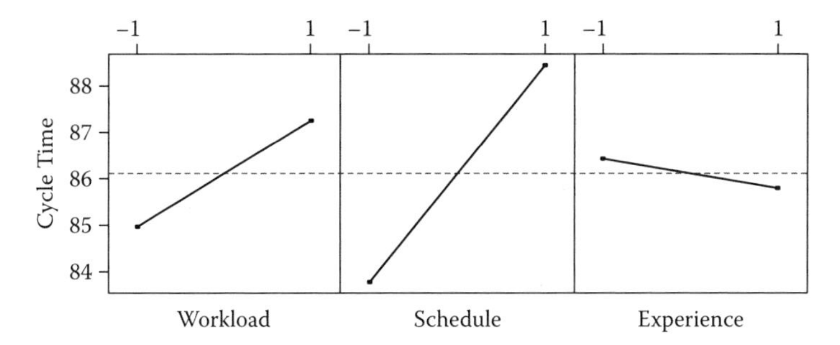

FIGURE 15.15 Main effects plot (data means) for cycle time.

if fully implemented, so it is important to understand what effect it would have before making that type of commitment. The second potential solution would be to reassign some of the current field superintendents with more experience to shorten cycle time. The third proposal was to change the communication system between the field personnel and the contractors. It was proposed that changing to a direct-connect cell phone network would increase the ability to communicate. The fourth proposal was to implement a new scheduling system that claims to reduce lead times and to allow for better management of resources throughout the process.

Given the resources available, the team could test only three of the four solutions proposed so a paired comparison analysis was done to select the top three solutions to test. The team decided to eliminate the new phone network variable and to test the remaining in a factorial design with 10 replications due to the amount of variation in the data:

- Superintendent load (Levels: low and high)
- Superintendent experience (Levels: less than five years and five or more years)
- Scheduling system (Levels: original and alternate)

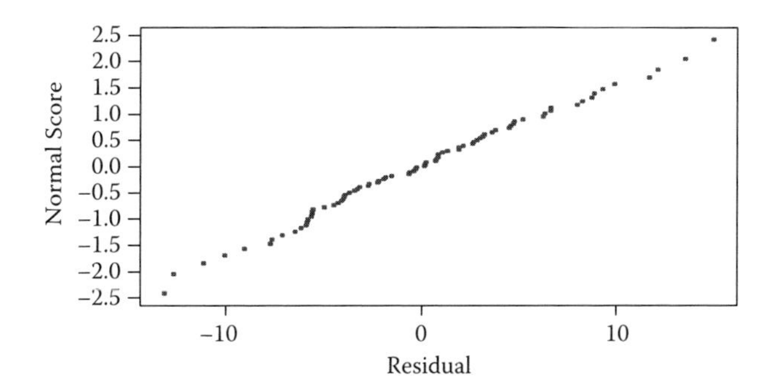

FIGURE 15.16 Normal probability plot of the residuals (response is cycle time).

The following ANOVA table resulted from analysis of the experimental data.

Analysis of Variance for Cycle Time (Coded Units)

Source	DF	Sequence SS	Adjusted SS	Adjusted MS	*F*	*p*
Main effects	3	551.44	551.44	183.813	4.61	.005
Two-way interactions	3	17.78	17.78	5.927	0.15	.930
Three-way interactions	1	15.00	15.00	14.995	0.38	.542
Residual error	72	2871.52	2871.52	39.882		
Pure error	72	2871.52	2871.52	39.882		
Total	79	3455.73				

The analysis revealed a significant effect for at least one of the variables tested. The interactions between the variables tested were not statistically significant. The team next reviewed the corresponding table of main effects and interactions estimated.

Term	Effect	Coefficient	SE Coefficient	*t*	*p*
Constant		86.106	0.7061	121.95	.000
Workload	2.2979	1.1489	0.7061	2.42	.038
Schedule	4.6781	2.3390	0.7061	3.31	.001
Experience	–0.6383	–0.3192	0.7061	–0.45	.653
Workload*Schedule	0.1052	0.0526	0.7061	0.07	.941
Workload*Experience	–0.8058	–0.4029	0.7061	–0.57	.570
Schedule*Experience	0.4781	0.2390	0.7061	0.34	.736
Workload*Schedule*Experience	–0.8659	–0.4329	`0.7061	–0.61	.542

The data suggest that both superintendent workload and schedule had a significant impact on overall cycle time. Lower workloads lead to lower cycle times, and the alternate scheduling system also leads to lower cycle times. The main effects plot for each independent variable (IV) is displayed in Figure 15.15.

Review of the diagnostic plots yielded no significant violations of ANOVA assumptions. The team recommended that the new scheduling system be adopted first and, if further improvement was still needed, then the workload of the superintendents be adjusted by reducing the number of homes for which the team is responsible at any given time. The normal probability plot in Figure 15.16 illustrates the associated diagnostic test.

Comprehensive Glossary of Continuous Improvement Terms

The following terms are used in the fields of continuous improvement, Total Quality Management, Lean, and Six Sigma. They are included as reference material.

Acceptance sampling: Evaluation of a portion of a lot for the purpose of accepting or rejecting the entire lot as either conforming or not conforming to a quality specification.

Action plan: The steps a team develops to implement a solution or the actions needed to make continued progress toward a solution.

Activities: Segments of programs that, when coordinated, add up to a particular program.

Adaptable process: A process capable of being changed to accommodate future business requirements.

Assignable cause: The name for the source of variability in a process that is not due to chance and therefore can be identified and eliminated. (See *Special cause*)

Attributes data: Data obtained by counting either nonconforming items or the occurrences of nonconformances. Also known as discrete data.

Brainstorming: An idea-generating technique that uses group interaction to generate many ideas in a short time period.

Breakthrough: A change; a dynamic, decisive movement to new, higher levels of performance. Used by J.M. Juran to describe a new, improved level of sustained quality performance.

C-chart: Control chart for plotting data based on the total number of nonconformances (defects) in a sample.

Cause: A proven reason for the existence of a defect.

Cause and effect diagram: A structured form of brainstorming that graphically shows the relationship of causes and subcauses to an identified effect (problem). (See *Fishbone diagram*)

Chance cause: The name for the source of variability in a process that occurs randomly. (See *Common cause*)

Charter: A commitment by management in document form stating the scope of authority for an improvement group.

Checklist: A sequential list of items to be attended to or of steps to be taken.

Check sheet: A tool for gathering information about a problem and its probable causes by collecting and organizing two or more kinds of information at the same time.

Chronic problem: A long-standing adverse situation that requires solution by changing either the process or the status quo.

Code of cooperation: A list of actions/behaviors agreed to by the team that fosters cooperative team interactions and effective team decisions.

Common cause: A cause of variation in a process that is random and uncontrollable. (See *Chance cause*)

Continuous data: Data that are generated by measuring. Also known as variables data.

Consensus decision: A decision made after all aspects of an issue, both positive and negative, have been brought out to the extent that everyone openly understands and supports the decision and the reasons for making it.

Continuous improvement: Continuously monitoring processes to determine if they function as desired and if they can be improved.

Control chart: A chart showing sequential or time-related performance of a process that is used to determine when the process is operating in or out of statistical control using control limits defined on the chart.

Control limits: A statistically derived limit for a process that indicates the spread of variation attributable to chance variation in the process. Control limits are based on averages.

Controllability study: A study to learn if defects are operator controllable or management controllable.

Cost of quality: The cost of conformance (achieving quality) plus the cost of nonconformance (waste).

Critical dependencies: The interrelationships existing within or among processes that are primary drivers of defects or errors in a product or service.

Critical success factors: Indicators developed by a customer that indicate the defect-free character of a product or service.

Customer: Anyone for whom an organization provides goods or services.

Defect: A nonconforming attribute.

Defective: A unit with a nonconforming attribute.

Deming, W. Edwards: Father of the "Third Wave" of the industrial revolution in Japan; advocates quality and productivity improvement through statistical process control.

Deming Wheel: Plan, Do, Check, Act (PDCA): To achieve quality improvement, Deming says you must plan for it, implement it (do), analyze the results (check), and take action (act) for continuous improvement.

Department task analysis: A method for analyzing an organization by determining its mission and how it interacts with customers and suppliers. Used to position the organization for improvement.

Diagnosis: The process of studying symptoms, taking and analyzing data, conducting experiments to test theories, and establishing relationships between causes and effects.

Diagnostic arm: Term used by Juran to refer to a person or persons brought together to support data gathering and problem analysis.

Effect: An observable action or evidence of a problem.

Effective: A process that delivers a defect-free product or service to the customer.

Efficient: A process that operates effectively while consuming the minimum amount of resources (e.g., labor, time).

Errors, inadvertent: Workers' errors that are unintentional, unwitting, and unpredictable.

Errors, technique: Errors that arise because workers lack an essential technique, skill, or the knowledge needed to avoid making the error.

Errors, willful: Errors that workers know they are making.

Facilitator: A person who functions as the coach/consultant to a group, team, or organization. In quality improvement, the facilitator focuses on process while the team leader focuses on content.

Failure, accidental: Failure arising from misuse while in service.

Failure, infant mortality: Early service life failure due to misapplication, design weaknesses, manufacturing mistakes, or shipping damage.

Failure, wearout: Failure after acceptable service life.

Fishbone diagram: Another name for the cause and effect diagram. (The finished product resembles a fish skeleton. Also called an *Ishikawa diagram* after the Japanese engineer who first developed it.)

Fitness for use: The condition of goods and services that meets the needs of people who use them.

Flowchart: A chart that symbolically shows the input from suppliers, the sequential work activities, and the output to the customer.

Force field analysis: A list identifying promoting and inhibiting factors (forces) that must be overcome before opportunity/problem lists can be built or effective solutions can be implemented.

5 Ws and 1 H: Who, what, where, when, why, and how. A useful tool to help develop an objective and a concise statement of the problem.

Gatekeepers: Individuals who help others enter into a discussion (gate openers) and those who cut off others or interrupt them (gate closers).

Goal: A statement describing a desired future condition or change, without being specific about how much and when.

Histogram: A bar chart that illustrates the frequency distribution of a measurement or value.

Imagineering: Visualizing what a process without waste can become as a goal for improvement activities.

Impact changeability analysis: A tool for prioritizing a list of problems/opportunities by ranking them according to the degree of impact versus the ease of change.

Implementers: Individuals responsible for performing tasks within a process.

In control: A process operating with assignable causes of variation (or special causes) is said to be "in a state of statistical control," which is usually shortened to "in control."

Intervention: The role of a team facilitator when he or she interrupts a group to state his or her observations about the group dynamics.

Ishikawa diagram: Another name for a cause and effect diagram, named after the engineer who developed it. (Also known as a *fishbone diagram*.)

Juran, Joseph: Emphasizes management's role in quality improvement by solving chronic problems, project by project.

Just-in-time (JIT) inventory: The minimum inventory required to meet production schedules.

Management-controllable defect: A defect that does not meet all of the criteria for an operator-controllable defect. (See *Operator-controllable defect*)

Matrix concept: A group of elements with rows and columns designed to cross-reference multiple measurements or sets of data.

Measurement: The dimension, quantity, or capacity determined by measuring.

Meeting assessment: A process where the team collects information about the effectiveness of their meeting. (See *Process check*)

Mission: The single overriding goal statement for an organization. It should encompass all organized activities that are significant in terms of resources used.

Multivoting: A structured series of votes by a team that reduces a list containing a large number of items to a manageable few.

Natural tolerance limits: A three-standard-deviation (3-sigma) spread, both above and below the mean, in the distribution of individual occurrences of the process. The natural tolerance limits are not necessarily related to the specification limits but reveal the natural random variability of a process.

Nominal group technique (NGT): A technique for generating a large number of ideas in a short period of time. NGT differs from brainstorming in that team members are asked to prepare a list of ideas prior to the session.

Normal distribution: A continuous, symmetrical, bell-shaped frequency distribution for variables data.

Np-chart: Control chart for plotting the number of nonconformances in a sample.

Number defective: Total number of defective units found in a sample.

Number of defects: Total number of defects found in a sample.

Objective: A more specific statement of the desired future condition or change than a goal. It includes measurable end results to be accomplished within specified time limits.

Operational definition: A way to define something in observable/measurable terms.

Operator-controllable defect: A defect that occurs where it is possible for operators to meet quality standards. It is controllable when operators know what is expected, know what their actual performance is, and have a means for regulation.

Out of control: A process with variation outside the control limits.

P-chart control chart: Used to evaluate performance based on the percent of products with nonconformances (percent defective).

Pareto diagram: A type of bar chart prioritized in descending order from the left to right, distinguished by a cumulative percentage line that identifies the vital/critical few opportunities for improvement.

Problem statement: A statement that describes in specific, concrete, and measurable terms what is wrong and the impact.

Problem-solving process: A flowchart of specific tasks by which chronic system problems can be solved.

Process: A group of usually sequential, logically related tasks that use organizational resources to provide a product or a service to internal or external customers.

Process capability analysis: A statistical technique used during development and production cycles to analyze the variability of a process relative to product specifications.

Process check: An assessment of the meeting process versus content.

Process flowchart: (See *Flowchart*)

Process owner: A manager assigned responsibility for the quality improvement of a process and given authority to make improvement happen.

Process quality improvement: A comprehensive method for describing and analyzing an organization's processes that sets the stage for effectively monitoring and controlling these processes.

Process rating: The result of evaluating a process against criteria relating to its effectiveness, efficiency, and adaptability. The rating of a given process is improved through application of process quality improvement.

Program: A planned, organized effort directed at accomplishing an objective. A program specifies how an objective is to be reached.

Quality: Quality is providing customers with products and services that consistently meet their needs and expectations.

Quality audit: An independent evaluation of various aspects of quality performance.

Quality circle: A group of people from the same workgroup who focus attention on ideas for improving quality within their own area.

Quality control: The process of measuring quality performance, comparing it with the standard, and acting on the difference.

Quality costs: (See *Cost of quality*)

Quality Improvement Center (QIC): Mission is to establish a resource to assist and support management in implementing the continuous improvement of quality and productivity.

Quality improvement: A systematic method for improving processes to better meet customer needs and expectations.

Quality improvement experts: (See *Deming, Ishikawa, Juran,* and *Taguchi*)

Quality improvement team: A group of people who meet to identify, analyze, and solve chronic system problems/opportunities.

Quality planning: Launching, for example, new products or processes in which continuous quality improvement is built in.

R-chart: A control chart of the range of variables as a function of time, lot number, or similar chronological variable.

Random cause: (See *Common cause* or *Chance cause*)

Random sample: The number of units chosen from a lot by a method that gives each unit an equal chance of being selected.

Recorder: The person who takes minutes for meetings.

Remedy: (See *Solution*)

Rework: To correct defects a process has produced.

Robust design: A design of a product, service, or process that is insensitive (robust) to changes in uncontrollable (noise) variables.

Root cause: The basic reason creating an undesired condition or problem. In many cases, the root cause may consist of several smaller causes.

Run chart: A graphic plot of a measurable characteristic of a process versus time.

Sample size: The number of units to be selected for random samples.

Scrap: The loss in labor and materials resulting from defects that cannot economically be repaired or used.

Scribe: A person who records inputs from a team on a pad or board.

Solution: A change that can successfully eliminate or neutralize a cause of defects.

Special cause: A cause of variation in a process that is not a random or uncontrollable cause, in contract to a common cause.

Specification limits: Limits established for a process or a product that are determined by engineering, development, or the customer. Specification limits are applied to individual occurrences and are not related to natural tolerance limits.

Sporadic problem: A sudden adverse change in status quo requiring a solution that returns the condition to the original state.

Stable process: A process in statistical control.

Standard deviation: A mathematical term to express the variability in a data set or process.

Statistical process control (SPC): The application of statistical methods to monitor variation in process *inputs* over time. SPC displays variation in process inputs to identify special/assignable causes of variation.

Statistical quality control (SQC): The application of statistical methods to monitor variation in process *outputs* over time. SQC displays variation in process outputs to identify special/assignable causes of variation.

Steering arm: A term used by Juran to refer to a person or persons from various departments who give direction and advice on an improvement program.

STP analysis: Situation, target, potential solution. A method to organize information (perceptions) about a problem into three categories: the way things are now (situation), what the team wants to accomplish (target), and information about ways to get from the current situation to the desired target or outcome (potential solution).

Structure tree diagram: A visual technique for breaking a problem into its component parts. The starting point is a general statement, and the branches of the tree are formed successively, breaking down the general statement into more specific statements.

Subprocess: A group of tasks that together accomplish a significant portion of an overall process.

Supplier: Anyone from whom the organization receives goods or services.

Symptom: A condition where evidence of a problem is manifested.

Systems audit: An evaluation of any activity that can affect final product quality.

Taguchi, G.: Stresses optimizing design specifications. Popularized concept of robust design.

Task: Specific activity necessary in the function of an organization.

Team leader: A person who leads a team through the problem-solving process.

Team member: A person trained in identifying, analyzing, and solving chronic system problems and identifying improvement opportunities.

3 Ps: Purpose, process, payoff. Key areas to discuss at the start of a meeting or presentation are the purpose of the meeting, the process that will be followed, and the payoff to the group.

Theory: An unproven assertion about the reasons for the existence of defects and symptoms.

Total quality management (TQM): Originally conceptualized by Feigenbaun, is a management philosophy that focuses on the need to continuously improve the quality of a company's products and services.

Trend: A gradual change in a process or product value away from a relatively constant average.

U-chart: Control chart for plotting data based on the number of nonconformances in each unit (defects per unit).

Variability reduction process (VRP): Begins with the voice of the customer, which is dissected and deployed throughout the producer organization using the multiple matrices of quality function deployment (QFD), then analyzed using design of experiments (DOE), and finally monitored using statistical quality control (SQC) in the early stages of the variability reduction process and statistical process control (SPC) in the later or more mature stages of the VRP.

Variables data: Data resulting from quantitative measurements. Also known as continuous data.

Variation: A concept that no two different items will be completely the same.

Voice of the customer (VOC): A set of the collective verbatims provided by or solicited from actual and potential customers to better focus on product or service design.

Voice of the customer table (VOCT): A matrix-like table used to translate customer verbatims into positive, useful statements of what customers expect or are looking for in the products and services they desire.

Waste: Anything we expend resources on that does not add value to the final product.

$\overline{X}$: The arithmetic average of a data set composed of two or more values.

$\overline{\overline{X}}$: The arithmetic average of a collection of $\overline{X}$ values. Referred to as the average of the averages.

$\overline{X}$–R chart: A variables control chart including $\overline{X}$ to track the process average and R to track process variability (called an X-bar and R chart).

Pertinent Web Sites Addressing Continuous Improvement

http://en.wikipedia.org/wiki/Continuous_improvement
http://en.wikipedia.org/wiki/Kaizen
http://www.asq.org/learn-about-quality/continuous-improvement/overview/overview.html
http://managementhelp.org/quality/cont_imp/cont_imp.htm
http://www.1000ventures.com/business_guide/mgmt_kaizen_main.html
http://www.west.asu.edu/tqteam/other.htm
http://systems2win.com/
http://www.gembapantarei.com/2009/03/accountability_for_continuous_improvement_1.html
http://www.grokdotcom.com/2008/08/28/what-is-continuous-improvement/
http://www.leansupermarket.com/servlet/StoreFront?gclid=CLiU2LjQ9JkCFRlcagodLiigSA
http://www.quality-control-plan.com/qualitymanagementsystem-kits.htm
http://www.qualproinc.com/
http://www.industryweek.com/articles/continuous_improvement_--_a_better_way_of_developing_new_products_16321.aspx
http://www.statsoft.com/textbook/stathome.html

Index

W

X

Z